The Subtle Beast

The Subtle Beast

Snakes, from Myth to Medicine

André Ménez

CRC Press
Taylor & Francis Group
Boca Raton London New York

CRC Press is an imprint of the
Taylor & Francis Group, an **informa** business

CRC Press
Taylor & Francis Group
6000 Broken Sound Parkway NW, Suite 300
Boca Raton, FL 33487-2742

© 2003 by Taylor & Francis Group, LLC
CRC Press is an imprint of Taylor & Francis Group, an Informa business

No claim to original U.S. Government works

Visit the Taylor & Francis Web site at
http://www.taylorandfrancis.com

and the CRC Press Web site at
http://www.crcpress.com

Contents

Preface

This book introduces the reader to the complex and absorbing world of snakes. Each of the fourteen chapters can be read independently, but the reader will profit most by reading the book from cover to cover, since it traces a fascinating journey from the macroscopic features of snakes to the molecular description of their venom components. The first three chapters describe how, through the ages, snakes have inspired fear or love, have been cast as gods or devils, as living representations of good or evil, and even as fantastic medicinal tools. The sway of such supernatural beliefs has progressively eroded since the seventeenth century by a number of great scientists, whose work is described in the eleven remaining chapters. The first three of these look at the origin, features and classification of snakes, and then an increasingly precise description of snake venoms and their components leads to the final chapter, which explains how venoms are useful to us. On his or her journey through the book, the reader will bear witness as the snake sheds its mythical skin to emerge in the full splendour of its true self.

Acknowledgements

I wish to very warmly thank the following persons for their kind help, contribution or patience: Michael J. Benton, René Cado, Antonio Camargo, Cécile, Henriette Clévier, Christophe, Thierry Damerval, Jay Fox, André Galat, Dominique Hauteville, Hal Heatwole, Daniel Heuclin, Peter Houghton, Ivan Ineich, Florence Izabelle, Christian Jacq, Carlos Jared, Carlo Laj, Cynthia Lee, Francine Lefèvre, Jürg Meier, Angelo Moreto, Frédéric Mûller, Stefan Niewiaroswski, Danièle Patron, Jean-Claude Rage, Olivier Remondière, Philippe Savarain, Denis Servent, Reto Stöcklin, André Syrota, Jacques Thaler, Julian White, Wolfgang Wüster.

A special thank you to Jean-Philippe Chippaux, Max Goyffon, Manjunatha Kini and Nicolas Vidal for their critical reading of the manuscript. Special thanks also to Daniel Heuclin for his kind gift of so many photographs, and to Renée Ménez for her constant help and support. Thanks are also due to the authorities of the Commissariat à l'énergie Atomique for giving me permission to write this book.

Men and snakes: truths and fallacies

The good old times, when a well-rounded person might master all aspects of science, seem, alas, to have gone. Science has become such a vast universe, divided into so many apparently strictly separated areas, that it is becoming increasingly difficult to find one topic of interest to everybody, especially if the reader is not seeking a popularised approach.

Undoubtedly, zoology, the biological science of animals, is a privileged domain because everyone has some idea of what it is about. An agreeable discipline, it is usually welcomed by school children, so the basic elements of zoology are absorbed early in life. Moreover, we are surrounded by all sorts of animals, from pets to pests, offering us an immediate opportunity to appreciate and understand animal behaviour. Television, too, is a major source of information with many, often award-winning and popular, natural history documentaries.

Among the different creatures that populate the world around us, snakes occupy a special place; hardly anyone is indifferent to them. They are often beloved, sometimes excessively, or just hated. Most people are strikingly well informed about them; just ask anyone to describe a snake. Back will come the answer that snakes are long animals with no legs, able to crawl on the ground, swim in rivers (people are often unaware of true sea snakes) and climb trees. Most people know that snakes can be poisonous, although this is sometimes a source of confusion as many snakes are rightly recognised as not venomous. It is also common knowledge that snakes periodically shed their skin, from the tip of the snout to the end of the tail.

Perhaps even more surprisingly, the specialised names of some snakes are widely familiar. For instance, the one commonly called a 'viper' in Europe is a 'rattlesnake' in the United States. 'Cobras' and 'Najas' (Figure 1.1) mean types of snakes to most people, although they may not remember exactly what they look like. 'Pythons' and 'boas' are understood not to be venomous but are famous for their ability to coil around their prey, killing it by suffocation or heart failure. People know they are large, although their actual size is often exaggerated. Contrary to common belief, the anaconda is

Figure 1.1 A snake charmer in India (D. Heuclin).

not the largest snake, a title that belongs to the Indian python (*Python molurus*) which can grow to more than 10 m in length.

As we delve deeper, we are frequently confronted by another world, one of fantasies which are endlessly passed on and which often originate from ancient beliefs, legends, myths and even religions. It is fascinating to try to get to grips with what lies behind these misconceptions. Just think of a few common questions, often part of strongly held but misguided beliefs.

Snakes and worms
Are snakes related to worms? Definitely not. Probably because they look similar, and also because they slither, children sometimes confuse the two groups of animals. Interestingly, some historians of the last two centuries are convinced that the same confusion appeared in Bible stories. This is the case for the 'fiery serpents' which killed Hebrews crossing regions around the Red Sea on their long journey towards the Promised Land. Travelling along the valley of Arava, between Aqaba and the Dead Sea, on the rocky sea of Suph they met the fiery serpents sent by Yahweh to punish them for complaining of their misfortunes: 'And the Lord sent fiery serpents among the people, and they bit the people; and much people of Israel died' (Numbers 21:6).

Some historians interpret this story as a manifestation of *dracunculosis*, a disease caused by worms, not snakes. Freshwater lakes and marshes of certain tropical areas, especially in Africa and the Middle East, are known to be infested by filaria, a small endemic roundworm called *Dracunculus medinensis*. Their larvae are carried by a tiny sort of fresh-water shrimp, no more than a millimetre long, which infects people who drink contaminated water. It takes about a year of incubation for the worm larva to develop to the adult state. Usually only the disease-causing females can be detected. They are thin animals, 35 cm to 100 cm long (yes, up to a metre!), which migrate towards the outer tissues of the body, punching holes through the skin to release their larvae. Unfortunately, dracunculosis is still present in subtropical areas stretching from the west coast of Africa all the way to India. In 1758, Carolus Linnaeus, the famous Swedish naturalist and physician, gave a scientific description of the adult worm, and in 1871 Aleksej Pavlovich Fedschenko, a Russian zoologist, described its lifecycle.

If we interpret the biblical story of fiery serpents as a manifestation of dracunculosis, we have to assume that the chroniclers of the times confused snakes and worms. But is that really likely? The famous Egyptian medical papyrus discovered around 1870 by Georg Moritz Ebers, a German Egyptologist and novelist, described the disease as early as the fifteenth century BC, reporting that, when a human limb displays a sort of ulceration containing a larva (not yet identified as a worm), the skin should be pierced and the larvae removed gently with pincers. The long and thin whitish matter extracted looked like mouse brain! Plutarch, the famous Greek biographer, quoting Agatharcides, reported that those who travelled around the Red Sea suffered from a strange disease, with small snakes that came out of their bodies and ate their legs and arms. When touched, those snakes burrowed back into the sufferer's body and hid in their muscles, causing terrible pain. Clearly, a mystery has long surrounded the nature of the 'creature' that was responsible for dracunculosis. It was sometimes called the *dragonneau*, the *Pharaoh worm*, the *filaire of médine* or the *Guinea worm*. An interesting debate continues on whether or not snakes and worms were indeed confused in the past, particularly in the Bible. We shall most probably never know the answer.

Today there is no reason for confusion: snakes are not worms. They have a spinal column, or backbone, and so belong to the category of animals called 'vertebrates', a group including fish, amphibians (like frogs, toads and newts), reptiles, birds and mammals. We humans belong to the same broad group as snakes. By contrast, worms are invertebrates without any backbone.

Popular beliefs – most of them wrong
Are snakes really slimy? Quite commonly, even non-poisonous snakes repel people because they are considered to be slimy or 'gluey'. This is just not true. Reptiles are among the cleanest of animals and it is quite rare to find a 'dirty' snake. Why, then, are they thought to be slimy? Perhaps because of the smooth scales which cover their bodies, giving them a shiny appearance and allowing one to slip readily between your fingers if you try to pick it up.

Is a snake capable of hypnotising? Everyone has seen pictures of cobras watching small birds which seem totally paralysed. And we all remember the Walt Disney movie *The Jungle Book*, based on the famous collection of stories by Rudyard Kipling, which showed Kaa, a huge snake that subjugates Mowgli. However, this fascinating attitude of snakes should not be misinterpreted. Just look at a snake watching its prey. The snake is virtually expressionless – just a steady gaze suggesting profound concentration. Look closer and you will see that snakes have no eyelids and that the movements of their eyeballs are limited. Moreover, their pupils are large and dark, further adding to the impression of a fixed gaze. Snakes may seem to hypnotise their prey but it is most unlikely that they really do so.

Do snakes love music? Watch a snake charmer playing the flute: the slow movement of the snake, apparently following the musical beat, certainly makes it look as though the animal (usually a cobra chosen for its impressive posture) is an interested listener. But it probably cannot be true because snakes would find it difficult to hear music. While not deaf, their auditory apparatus, and therefore their hearing ability, is limited. Instead, they sense vibrations which reach them through the ground on which they lie and then they rise up, apparently having been woken by the music. Why then do they follow the beat? The snakes are simply trained to follow the movement of the flute in a defensive posture; if the flute was motionless, the cobra would sink back into the basket.

Do snakes really love milk? This belief was, and might well still be, strongly anchored in the minds of many pastoral societies in some countries in Europe and elsewhere. People thought that snakes could suck milk directly from cows, sheep and even from women sleeping near their babies. It was claimed that snakes are so attracted by milk that they approach babies in their cots and take it directly from the infants' mouths! Not unexpectedly, in times gone by, all sorts of defences were recommended to prevent snakes from approaching a house, a woman or a baby. At the beginning of the eighteenth century, it was thought that the best thing was to carry garlic, which, as some people believe, effectively repels demons. Alternatively, people were advised to carry the heart of a vulture, or wrap themselves in the leaves of an ash tree. Neither the milk-loving behaviour of snakes nor these repellent remedies have been confirmed by zoologists and you won't find any mention of them in textbooks!

Are 'minute snakes', as they are called in French literature, at all dangerous? Again the answer has to be 'no'. In fact, they got their name from the Latin *minutus* (small) because they are tiny, not because they could kill a man in a minute. These tiny reptiles, perhaps no more than 15 cm long, are totally harmless to humans (Figure 1.2). They use their heads to dig into the soil and some of them seem to be blind, hence they are called 'blind-snakes'. Blind snakes are harmless but, as we shall see later, there are many other snakes, powerful and dangerous, which can indeed kill a healthy man rather rapidly.

Are snakes' tongues poisonous? This is also a common but once more unfounded belief. Snakes' tongues are forked (Figure 1.3). They are pushed out through a groove in the front of their mouths, then flicker and briefly touch objects in their immediate vicinity, from which they collect chemicals enabling them to identify the surface they have just licked. The chemical molecules collected by the tongue, as well as those in the air, are carried to a specific organ in the nasal cavity which examines them and reports to the snake's brain. Tasting is a highly developed function in snakes which allows them to find out whether it is food or something less desirable that has just been tasted. The poison produced by venomous snakes has nothing to do with the tongue; it is injected by long, pointed teeth (fangs), linked to

Figure 1.2 Snake or worm? Just a blindsnake: *Typhlops vermicularis* from Turkey (D. Heuclin).

Figure 1.3 The tongue of an American crotal, *Crotalus viridis* (D. Heuclin).

specialised venom-producing glands usually located in the upper jaw behind the eye.

In some countries, including South Africa, there is a common belief that all snakes spit their venom. But once again this is another fallacy. Only a few snakes, such as the African cobra (*Naja nigricollis*) and the ringhals (*Hemachatus haemachatus*), spit. They eject a swift thin stream of venom more by squirting than spitting.

Why, at the very end of the twentieth century, are our minds still full of so many erroneous impressions? One possible explanation, but certainly not the only one, is the supernatural roles that snakes have played in human history, and particularly in mythology. Many ancient, fabulous and sacred tales, usually derived from remote religious beliefs, tell us how extraordinary deeds were accomplished by supernatural creatures which were often serpents. Passed down through the millennia from generation to generation, some stories came to us in a confused mixture of truth and legend.

That snakes played fabulous roles in mythology is not surprising. These creatures do such strange things: they move around relatively quickly without legs, manoeuvre well in trees and in water, swallow their prey whole, regularly renew their skin and sometimes inject a highly dangerous poison when they bite. Snakes certainly appear to be so very different, both from other animals and from ourselves, that it is easy to see why, until the relatively recent advent of rigorous scientific investigation, these reptiles were regarded almost as supernatural creatures.

2

Snakes and myths

Scientific books on snakes frequently include a chapter describing the part they have played in myths. The descriptions often correspond to ancient poetic visions; they remind us that many (and often excessive) virtues have been ascribed to serpents: strength, power, beauty, cleverness, nimbleness, a highly developed instinct, nobility and an ability to cause death. Snakes in stories were described as supernatural and, in myths, were thought to be devils or gods.

Being sometimes poisonous, hidden in the shadows, slowly and mutely gliding, snakes have often been deemed powerful and shifty, evil creatures whose major aim was to frustrate the natural and proper development of life. Sages speculated endlessly about the negative impact of serpents on such essential matters as the creation of the universe, the fate of human beings and of their gods. Just take a look at a few examples of how the literature through the ages has portrayed snakes and serpents.

In the Bible, the Devil assumed the form of a serpent and successfully persuaded Eve to pick fruit from the tree of knowledge of good and evil, the only tree in the Garden of Eden which God had strictly forbidden her to approach. Eve's action resulted in the loss of the privileges God had given to people: their punishment was the obligation to work, to beget offspring and to die. Snakes also had to pay the price for their imposture; they were condemned forever to crawl and, for those with a biblical tradition, to be the symbol of vice and lewdness.

The word *Ouroboros* is of Greek origin. It is the image of the snake that bites its own tail, so adopting a circular shape. It symbolises the perpetual continuity of life and death, both fertilising and swallowing itself. Following that idea, some cultures use similar words for 'snake' and 'life'. In modern Arabic, for example, life is *el-hayat*, and snake is *el-hayyah*. The circular shape of the Ouroboros also suggests a union between sky and earth, Nature appearing thus in its homogeneous plenitude.

Hesiod, a Greek poet of the eighth century BC, described the creation of the universe in some detail. Kronos, a primordial god and a son of

Ouranos (the sky) and Gaia (the earth), mutilated his father; from his blood emerged the Furies, the guardians of human life who hunted sinners. The Furies were awful to look at, with vipers for hair. All primordial monsters present during the creation of the world were then banned except for the Furies who will stay with us as long as there are sins on earth.

The ancient Egyptians also believed that the snake Atoum gave the day to the gods who then created the air and the earth. Being so important, universal and powerful, the gold cobra Uraeus naturally became the symbol of Egyptian sovereignty, life and knowledge (Figure 2.1). It was displayed as the sacred emblem on the foreheads of goddesses and members of the Egyptian royal family. That family ultimately vanished with Cleopatra, who was probably killed by a snake! In the same tradition, Ra the sun-god, vanishing each

Figure 2.1 Sethi the First with the gold cobra Uraeus, the symbol of Egyptian sovereignty.

day at twilight, penetrated the world of night and death, a valley occupied by the huge snake Apep, also called Apophis by the Greeks. Ra had to pass through twelve doors; Apep, continuously growing longer, opposed Ra's advance. But Apep always failed, thrust aside in the valley. Inevitably, every morning, the sun in its shining splendour appeared anew.

Many baleful creatures are said to have adopted a snake-like form, and popular heroes or gods often had to fight against them. Perhaps one of the most popular Greek myths is the life of Hercules, the son of Zeus and Alcmene, who performed the twelve labours. Pindar, a Greek Lyric poet of the fifth century BC, described how, when still an infant in his cot, Hercules grabbed two large, threatening snakes, one in each hand, and killed both. Later, during his second labour, Hercules moved to Lerne where he slew the Hydra, a nine-headed sea-serpent which grew two heads for each one lopped off. For his twelfth labour, Hercules defeated Cerberus, the guardian of the entrance to Hades, a three-headed dog represented with many snakes around his necks, whose bite is poisonous like that of a viper. For snakes and serpents to appear in three out of Hercules' twelve labours shows how important they were.

Ovid, the well-known Roman poet, told various snake-related stories about Perseus, the son of Zeus and Danae. With the help of Hermes, Perseus decapitated the mortal Medusa, one of the three Gorgons who had snakes for hair and eyes which, if looked into, turned the beholder to stone. Pegasus, the famous magical flying horse, was born from the blood of the Medusa; the Italian artist Caravaggio painted a remarkable picture of this terrifying creature (Figure 2.2). Perseus later rescued a beautiful young girl named Andromeda who was offered as a sacrifice to a horrible giant sea snake. As he did with Medusa, Perseus efficiently decapitated the monster. Later, he married Andromeda.

Snake-based representations of primordial forces, not necessarily always baleful ones, are, of course, not limited to Greek and Roman mythology. In India, for example, the giant snake Vritra retained waters and caused terrible droughts. Indra, the master of thunder, seasons and rain, opened Vritra's belly from which life-saving waters fell upon on the earth as rain or flowed as the seven rivers. Similarly, in Norse mythology, Midgard, the part of the world where men live, was thought to be a fortress encircled by a huge serpent. Thor, the god of thunder, caught and eventually killed the serpent of Midgard but was, in turn, killed by the reptile's poisonous breath.

Across the Atlantic in Mexico, snakes were also of central importance in religious rites; they can be seen in the remarkable sculptures of *Quetzalcoatl*, the feathered snake of the Toltecs. The snake was thought to represent earth, and the feathers spirituality.

Figure 2.2 The Medusa, the Gorgon with snakes for hair. Drawing from the painting by Michelangelo Amerighi, known as Caravaggio.

These stories show how important snakes have been in the traditions of many cultures; few other creatures have contributed so much to mythology. It is also significant that writers of a number of civilisations have thought that snakes symbolise both male and female characters. Their phallus-like shape and the capacity of some snakes to spit venom were seen as suggestive of male features, while their ability to coil up and embrace was thought to be female characteristics.

3

Snakes and early medicine

Have you ever thought about the 'caduceus', officially recognised as the emblem of the medical profession? If so, you may have noticed that it has a snake as a central component (Figure 3.1). The dictionary has two definitions. The first one describes it as a winged staff entwined with two snakes – this was the caduceus given by Apollo to Hermes, a terribly busy god who

Figure 3.1 The caduceus. Left: a caduceus used in the sixteenth century by Froben, a publisher from Basel (image kindly provided by Chris Mullen). Right: One of the various caduceus used nowadays by French medical doctors.

was known to the Romans as Mercury. Hermes served not only as the protec-
tor of travellers and rogues, as the conveyor of the dead to Hades and as the
messenger of the gods, but also as the god of inventions, commerce, theft
and many other things.

The second definition describes the caduceus as 'a similar staff to that
used as the symbol of the medical profession'. It is actually the emblem, or
more precisely the attribute, of Asclepius, the Greek god of medicine, who
was also the Roman Aesculapius. This form of the emblem is a single snake
twined around a rod or wand (Figure 3.2). Nowadays, the caduceus is considered to be the attribute of Hermes and Asclepius, and appears in either or
both its forms as the symbol of civilian and military medical and pharmaceutical professions in the US, the UK and other European countries.

Why does this organised duality of a snake and a staff represent medicine? The answer is a long and fascinating story which started several centuries before the birth of Christ and which, not surprisingly, contains several
mysteries. By and large, the story goes like this: in the ninth century BC,
Homer, the Greek epic poet, wrote in his famous *Iliad* that Asclepius was 'an
irreproachable doctor'. Various sources suggest that he was considered a sort
of hero in Thessaly: a man who could treat and cure diseases eventually
acquired a great reputation. His legend was born around the sixth century BC
and was reported by Pindar and a number of other poets.

Asclepius was then regarded as a demi-god, the son of Apollo – the
god of the sun, music, prophecy, poetry and medicine – and the mortal
Coronis. In the Tricca region of ancient Greece, Chiron, the wise centaur
who taught Achilles and Hercules, also tutored Asclepius in medicine. Asclepius naturally became the god of medicine, acquiring a universal reputation
for his ability to cure people with great efficacy. He was represented with a
wooden staff entwined with a snake.

The wooden staff symbolised not only the doctor who, at that time,
had to cover great distances to reach patients, but also trees; in the most
ancient Greek chthonian religion, trees were fundamental elements linking
the underworld with life on earth. In a sense, the staff symbolised the primordial forces of the world, emphasising that religion was also highly focused
on the gods and spirits of the underworld. Snakes were regarded as the
symbol of life, growth, rejuvenation, health, fertility and eternity.

It is possible that the tradition of a snake being so intimately wound
round the staff derives from the story of the Hebrews who, it has been suggested, suffered from filaria during their long journey toward the promised
land. The worms causing the disease (believed at that time to be snakes)
made holes in the skin as they emerged. One way of getting rid of the worms
is to roll them around a small stick and pull them out. Could the stick of

Figure 3.2 Asclepius, the 'irreproachable doctor', represented with a snake around a wooden staff.

Asclepius symbolise this procedure? This explanation was first suggested in 1674 but, in fact, nobody knows. However, as the Italians say: *Se non è vero, è bene trovato* (if it is not true, it is cleverly invented).

Asclepius was adored for his immense talent for efficient healing. Perhaps he was too efficient, because other gods clearly became jealous of him, considering his interventions too beneficial to humans. One day, Asclepius made a serious mistake by accepting a lavish honorarium to resuscitate a dead body. Zeus, the father of Apollo, was not pleased. How could a mere demi-god, half mortal, dare to be so powerful? Enraged, Zeus struck him down. After his death, Asclepius rose into the sky where, appropriately, he was transformed into the constellation of the Serpent.

Excavations in Greece at the beginning of the twentieth century revealed that the cult of Asclepius was celebrated in sanctuaries throughout the Peloponnese. The most famous and probably the richest of them was at Epidaurus, a sanctuary built during the fifth century BC, which became an important place of pilgrimage. The methods used to treat patients are unknown. They probably comprised of a blend of traditional and theologically-based medicines, divine intervention presumably operating whilst the patient was asleep in the sanctuary. Snakes were directly involved in the curative rites at the Asclepius sanctuaries. The reptiles were so effective that 'the blind could recover their sight simply by touching a snake tongue'! The curative efficacy of snakes gained popularity: in various cities, including Epidaurus and Rhodes, snakes were administered with great care and the rites associated with them persisted for several centuries. But just how the snakes contributed to the medical treatments remains a mystery.

The cult of Asclepius was observed in Greece for centuries and continued in Rome until 293 BC when the population there was decimated by a terrible epidemic of plague. As reported by Ovid, a poet born in 43 BC, the Roman senate sent an ambassador to Epidaurus, and Asclepius, the god of medicine, was 'imported' and became as popular in Rome as in Greece. Slaves who were cured in temples dedicated to Asclepius were freed on the order of the Roman emperor Claudius. And, as ever, Asclepius was depicted with his famous caduceus.

Despite the existence of multiple and powerful sects revering him, a mighty challenger emerged at the beginning of our era in the form of Christianity. The existence of a diety like Asclepius represented a considerable threat to the newly-emerging Christian Church, which accordingly struggled fiercely against this rival god. The Christians destroyed many sanctuaries and eventually the pagan worship of Asclepius vanished around the fifth–sixth century AD.

From the second to the fifteenth centuries, most alchemists found their inspiration in Hermeticus, a doctrine derived from the Egyptian god Thoth, the supposed author of works on magic, astrology and alchemy, and whose sacred bird was the ibis. The alchemists used the Hermes caduceus as the symbol of their art, which was essentially based on experiment. The Hermetic tradition vanished in its turn with the advent of the Renaissance.

At the time, most probably around the sixteenth century, the caduceus was again associated with the practice of medicine, with Protestantism contributing to its revival. As that religion developed, the saints of the Catholic Church were deliberately replaced by other well-known figures, Asclepius amongst them. This trend can readily be observed in medical books of that time which displayed these figures on their covers. The caduceus, often decorated with other features like a mirror, acquired its formal symbolism of the medical arts only at the end of the eighteenth century.

Since then, the caduceus, and especially the snake associated with it, has represented the deep wish of humans to cure those who suffer. Most strikingly, this pagan symbol, which has little to do with the original attribute of Asclepius, is gradually catching up with some contemporary realities. As we shall see later, snakes have recently been the source of a number of drugs which act effectively against various diseases. It took nearly 26 centuries for this astonishing connection to be made!

4

The origin of snakes

How snakes originated is an old, complex and poorly explained story based on a host of palaeontological studies. Snakes are reptiles; where did the reptiles come from?

We ought to start by sorting out the meaning of 'reptile', which has two possible connotations. Although it derives from the Latin *reptilis* (crawling), it is used formally in zoology to describe the class of *Reptilia* which currently embrace two groups of animals:

• the *Lepidosauria* include the tuataras (the lizard-like animals often called 'living fossils') and the squamates (lizards, snakes and amphisbaena also called 'worm-lizards' (Figure 4.1, Figure 4.2). Amphisbæna are limbless creatures with scales organised into regular rings. They live underground

Figure 4.1 A South American lizard, *Tupinambis nigropunctatus*, from Brazil (C. Jared).

Figure 4.2 An arboricolous iguana, *Brachylophus brevicephalus*, found on a small island called Kolonga, in Tonga.

in tropical areas and range in size between 10 cm and 75 cm (Figure 4.3). Their name comes from the Greek *amphis*, for 'both sides' and *bainein*, 'for walking'. These burrowing creatures can move forward or backward in tunnels with similar ease

- the *Archosauria*: the crocodilians (Figure 4.4) and the birds.

However, 'reptile' is also used for those early vertebrates, more correctly called 'amniotes', which appeared on land about 340 million years ago.

Primitive vertebrate life was aquatic; the oldest fossils suggest that fish appeared around 510 million years ago. To become terrestrial, aquatic creatures had to undergo extensive modification. The transition was slow, progressive and this new life became possible because essential features gradually developed: lungs to breath air, limbs to move on land, new skulls adapted to new feeding and breathing habits, new sensory systems, the maintenance of water balance against evaporation and an ability to reproduce on land.

The first of these changes can be detected in fossils of small animals called 'basal tetrapods' (i.e. having four legs). These ancient animals are often termed *amphibians* (meaning 'both lives') because they probably had a 'double lifestyle', just like frogs, newts and salamanders which constitute

Figure 4.3 An amphisbaena, *Amphisbaena alba*, from Brazil (C. Jared).

Figure 4.4 A caiman, *Caiman yacare*, from Mato Grosso in Brazil (C. Jared).

today's amphibia. It seems likely that basal tetrapods laid their eggs in water where the young hatched out as aquatic larvae, the tadpoles. As long as they are living in water, the tadpoles breath through gills and then metamorphose into an adult form for life on land.

The earliest tetrapods which moved onto land probably diversified into at least 40 different families, principally during the Carboniferous period 362–290 million years ago. Some palaeontology textbooks like to show an imaginative representation of East Kirkton in Scotland, around 340 million years ago. On the basis of the many fossils discovered there since 1830, the imaginary scene shows the valley with a rich flora, steaming hot springs at the base of active volcanoes, various aquatic arthropods and a sort of large salamander sitting on a rock. Indeed, a number of diversified tetrapods adopted clear reptile-like shapes; one sort were the aistopods, some of which were nearly 1 m long with 230 vertebrae. Various other reptiliomorph (i.e. 'reptile-bodied') tetrapods, like the diadectomorphs, lived in the period between 290 and 250 million years ago. They included rather large animals, up to 3 m long, with massive, short limbs and heavy vertebrae. Reptilian body structure seemed to provide an appropriate answer, at least in the early days, to the pressure for life to become terrestrial. While many primitive tetrapod forms became extinct, some evolved into the frogs, toads and sala-manders which still exist today.

Laying eggs on dry land

Some ancient amphibians underwent even more profound change, giving way to the earliest terrestrial vertebrates; as far as we know, these early animals appeared on earth not very much later than the original amphibians. Possibly, the oldest creature that was *not* an amphibian is the *Casineria*, a small reptiliomorph tetrapod about 15 cm in length with five fingers on each of its two forelimbs, recently discovered in East Kirkton; it may date from 330 million years ago. A little later, 310–300 million years ago, there were some small lizard-sized tetrapods, including the 20 cm *Hylonomus* and *Paleothyris*. They probably lived near ponds and lakes in luxuriant forests with tall trees. They all had a number of specific characteristics, one of which was truly remarkable for the time: their eggs. They laid eggs that did not need to hatch in water; their eggs were amniotic.

Amniotic eggs have two major characteristics;

- semi-permeable shells, usually though not always calcareous (that is, solidified with calcium salts like chalk), allow gas exchanges: oxygen can enter, and the waste carbon dioxide can leave, but the fluids remain inside. The fluids are essential for the embryo to develop into a terrestrial hatchling; they replace the water in which these animals' aquatic ancestors laid their eggs;

- the eggs have extra membranes outside the embryo which are essential for embryo protection, gas transfer and storage of waste elements. This was perhaps *the* major, adaptation to terrestrial life.

What happened next?

Ancestral snakes

The first amniotes gave way to three lineages, the *anapsida*, the *synapsida* and the *diapsida*, which can be distinguished by the presence or absence of *temporal fenestrae*, openings in the cranial box behind the eye orbits, which reduce the weight of the skull and store calcium (Figure 4.5). The *anapsida* has no such fenestrae; the *synapsida* has one; and the *diapsida* has two.

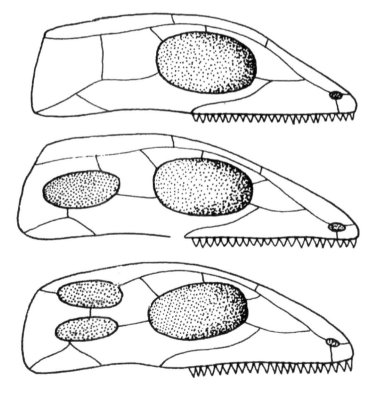

Figure 4.5 The skulls of anapsids (top), synapsids (middle) and diapsids (bottom), with respectively no, one and two temporal fenestrae. Redrawn with permission from M.J. Benton (1997).

Snakes are derived from the diapsids (Figure 4.6). Possibly the oldest example, called *Petrolacosaurus kansensis,* lived some 300 million years ago and was about 40 cm long. The diapsids developed well, especially during the Triassic period (250–205 million years ago) following the largest extinction of all time, resulting from a major disturbance to the environment.

They were replaced by the famous dinosaurs which spread all over the globe during the Jurassic period, in particular about 170 million years ago; the name *dinosaur* literally means 'terrible lizard'. Most of these fantastic animals disappeared mysteriously 65–70 million years ago. The diapsids left

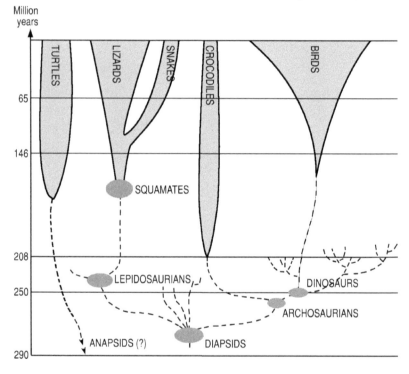

Figure 4.6 Evolution of amniotes. Only animal groups with presently living descendants are shown. Many other extinct groups populate this phylogenetic tree but they have been discarded for sake of clarity (for example, the dinosaurs are missing). These creatures lived during the Jurassic (208–146 My ago) and the Cretaceous (146 to 68 My ago) periods. Snakes arose from lizards. The question mark associated with the branch that led to turtles indicates that a recent genetic analysis has suggested that turtles may be quite close to crocodilians and birds and hence may not be the living descendants of anapsids, as is currently portrayed in traditional textbooks. Redrawn with permission from M.J. Benton (1997).

a few descendants whose present survivors are categorised as the *Archosaurians* and *Lepidosaurians*, the two categories of familiar animals we looked at a few pages back (pp. 19–20).

Recent studies of reptile genes have suggested that squamates emerged about 245 million years ago. Some lizards may have 'lost' their legs more than 95 million years ago as suggested by the discovery in the Middle East of the fossil of *Pachyrhachis problematicus* (what a name!), a 1m long animal with a snake-like shape and reduced rear legs. Snakes, therefore, arose from lizards.

A taste of evolution

The evolution of living organisms, snakes among them, encompasses all the changes undergone by them as their descendants develop into different living beings. A century or two ago, this phenomenon seemed to depend on two prime factors: life and time. How and why did these changes take place?

Georges Cuvier and Jean-Baptiste de Lamarck in France, and Charles Darwin in England, proposed three different theories, all assuming that external factors influence biological evolution. Cuvier was a creationist who propounded a catastrophe-based theory. For him, the species remained unchanged but were regularly exterminated and replaced by new ones. This interesting proposal fitted nicely with the observation that the limits of geological strata apparently reflect a renewal of the fauna: throughout the earth's history, new geology meant new animals and plants.

Lamarck was one of the first evolutionists. He believed that living organisms are adapted to (though the word 'adapted' was not yet used) and influenced by their environment. In his remarkable book, *Philosophie Zoologique*, he summarised the current classification of living organisms and claimed that time is one of the factors that is associated with their progressive transformation.

At the heart of Darwin's theory, so wonderfully described in his famous book *On the Origin of Species*, is the notion of species variability also proposed around the same time by his friend Alfred Wallace. Darwin's idea was based on the observation that all individuals within a population differ to some extent from one another, reflecting the variability that occurs in all species. Just look at the people around you. In broad terms we all look alike but in detail we are all different: tall and short, dark and fair, fat and thin, healthy or with this illness or that – only with identical twins is there any real difficulty in distinguishing between individuals. Our differences, and those of all our living neighbours, may be advantageous or disadvantageous in life's competitive struggles.

Darwin's second key idea was 'natural selection'. Through slight differences, some individuals may be more favourably adapted than others to their environment and hence will have a greater chance of leaving more offspring. After many generations, the offspring of those better adapted individuals would dominate in the population. *Individuals* do not evolve; it is the *species* which undergoes evolution because some individuals with particular characteristics leave more descendants than others.

The great achievement of Darwin's theory is that it provides a plausible explanation for the formation of species within the general framework of biology, something achieved long before the genetic material of organisms was discovered. Darwin's proposal essentially resulted from his remarkable and careful observation of individual living creatures. It was not easy to see the very slow processes of evolution actually at work but Darwin nevertheless concluded that, over time, different species diverge in their development, suggesting that once they shared a common ancestor; the more closely related the two species were, the more recent in their histories was that ancestor. This concept is, of course, now very widely accepted, with the ancestry of today's living organisms often being represented by a sort of family (or phylogenetic) tree to depict genealogical evolution.

Remarkable though it was, Darwin's theory did not explain everything. The concept of natural selection among animals was attractive, but many details awaited clarification. Moreover, while Darwin provided evidence that a 'transformation mechanism' occurs within a species which serves as the origin of other species, he did not demonstrate that the general evolutionary process on a geological timescale results from such a mechanism. Nor was he able to identify the 'motor of the transformation' as the genes and their mutations. Nonetheless, and perhaps most importantly, his work did provide a general understanding of biological evolution; it was and still remains a formidable inspiration for further studies associated with such illustrious names as Hugo de Vries in Holland, Julian Huxley and Richard Dawkins in England and Stephen Jay Gould and Ernst Mayr in the USA. Scientific progress in the field of evolution is of great general interest, as illustrated by the successes of many works on the subject. Who has not read – or at least heard of – '*The Panda's Thumb*' by Gould or Dawkins's book, '*The Selfish Gene*'?

A modern view of evolution

One interesting recent book on evolution is *Les horloges du vivant* by Jean Chaline; published in France, it shows just how complex evolutionary theory has become. Living organisms are based on a complex set of hierarchies, from the simplest at the level of biochemical molecules, upwards through genes and chromosomes, cells and organs to individuals and eventually to species, societies and finally the whole biosphere. Events at each of these levels can affect evolution. Major motors of change certainly include gene mutations which take place by substitution of one or more of the DNA bases or replacement of larger gene segments.

It is increasingly understood how gene modifications may cause gradual changes within a species or sometimes occur more abruptly, resulting in evolutionary discontinuities, jumps in evolutionary development if you prefer. In particular, recent molecular studies have suggested how jumps can occur within living organisms. An abundant anti-stress protein (HSP90) may play a crucial role. As well as binding to and thus protecting unfolded proteins, including a number of receptors, HSP90 may have an additional function. Two American scientists, Suzanne Rutherford and Susan Lindquist, have proposed that HSP90 might suppress the effects of mutations occurring during embryo development. Were a gene in an embryo coding for an essential protein to mutate, the consequences for the resulting protein could be compensated for by HSP90. Thus, potentially important genetic mutations might take place in a population without causing any structural change in the individuals and hence without being acted upon by natural selection. However if, during embryo development, HSP90 proteins were tied up elsewhere, perhaps as a result of a stress, the 'silent mutations' might be expressed and new characters, possibly including structural ones, emerge in a few descendants. In a sense, HSP90 would be acting as an evolutionary buffer. This interesting idea might account for rather slow evolution over long periods of time, followed by rapid transformations during a crisis.

Geological events throughout the earth's history are major contributors to evolution, for instance by cataclysms causing extinction of some populations thereby leaving room and opportunity for the development of new forms of life. Evolution therefore appears essentially chaotic because of the multiplicity of these underlying factors and, like the weather, is very hard to predict in the long term.

5

What are snakes?

Snakes, as defined in the dictionary, are legless, scaly reptiles with a long tapering cylindrical body belonging to the suborder of *Ophidians*. Yet snakes represent much more than this simple morphological definition. Through their remarkable diversity, their astonishing behavioural attitudes, their striking capacity to survive in varied, sometimes hostile environments and the ability of some to produce venom, snakes constitute a world to which many biologists have devoted their lives.

With a backbone, circulatory and nervous systems, an efficient digestive apparatus like most carnivorous animals, a respiratory apparatus (just a single long lung), a reproductive system, kidneys, muscles and so on, snakes have all the characteristics of terrestrial vertebrates. However, this is not the place to catalogue snakes and their characteristics, but rather to take a wider view of snakes in general.

The 'basic snake'

The movement of snakes looks magical. Though they have no legs, snakes can move fast and quite efficiently in all sorts of situations. Their long backbones comprise of between 100 and 400 vertebrae and, with a very special arrangement of muscles, they can adopt several efficient modes of locomotion and, furthermore, can switch virtually instantaneously from one to another.

The most common is the serpentine movement used by most legless reptiles when they are chasing prey or swimming (Figure 5.1). During this elegant side-to-side undulation, each part of the body passes across the same spot on the surface. There is also the concertina-type movement (Figure 5.2) consisting of anchoring one part of the body on the substratum while the rest of it is either pulled or pushed along, clearly an efficient strategy when moving through narrow tunnels.

The sidewinding motion (Figure 5.3) is used on moderately rugged surfaces, including unstable sand dunes. This motion consists of the propagation of an asymmetrical undulation together with a vertical bending that allows

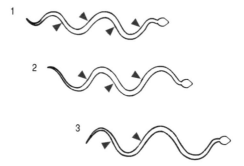

Figure 5.1 In lateral serpentine undulatory crawling, the whole ventral surface of the body touches the ground. The arrows show the main areas where pushing of the body occurs. Adapted from Ernst and Zug in *Snakes in Question*, published by the Smithsonian Institution, 1996.

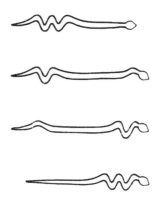

Figure 5.2 In the concertina movement, the extended part of the body slides over the surface. Adapted from Ernst and Zug in *Snakes in Question*, published by the Smithsonian Institution, 1996.

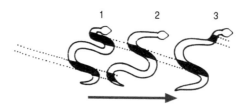

Figure 5.3 The sidewinding movement is useful to limit the contact of the body with the ground (dark regions). Adapted from Ernst and Zug in *Snakes in Question*, published by the Smithsonian Institution, 1996.

the mobile parts of the body to be raised. This is probably advantageous when moving on very hot land because the snake's body has little contact with the ground. There is also the caterpillar-like movement which allows snakes to move slowly in a fairly straight line, mostly by anchoring the ventral (belly) scales and successively extending and compressing the body. This mode of locomotion is often observed when tree snakes have difficulty moving along small branches. Finally, there is the fairly inefficient dragging thrust that snakes adopt on slippery surfaces or when facing an emergency. They anchor their head and from this point generate a large undulation which pushes the body laterally. When the wave reaches the tail, another symmetrical undulation starts, resulting in an overall sideways movement.

Have you ever seen the elegant swimming motion of a sea snake in a coral reef? The shape of the snake's body and its eel-like ability to undulate allow it to swim highly efficiently. Most sea snakes also have a paddle-shaped tail which is a morphological adaptation to marine life and helps their propulsive swimming.

Are snakes cold-blooded?
It is often claimed that snakes are 'cold-blooded'. More precisely, like other reptiles – and in contrast to mammals – their body temperature depends on the environment, so they are said to be *ectothermic*. They regulate their body temperature by exploiting various behavioural attitudes such as warming themselves in the sun. For instance, the yellow-bellied sea snakes (*Pelamis platurus*) remain so still in the sun at the surface of tropical seas that they can be approached quite closely provided, of course, that no noise is made. European vipers also sun themselves, often on top of a log which isolates them from the ground. Snakes even seek warmth at night and, in tropical regions, they often rest on tracks which have accumulated heat during the day. In excessive heat, as in some deserts, they cool themselves by burrowing into the sand, leaving their eyes just above the ground. They are also able to adjust their body temperature by living nocturnally or diurnally or, in cold conditions, by hibernating in deep holes.

Seeing
Snakes have classical senses: sight, smell, touch, probably taste (as suggested by the presence of taste buds in some snakes but not demonstrated in all) and hearing. With no movable eyelids, protection for their eyes is provided by a fixed transparent shield which is shed together with the whole skin during sloughing. Though vision in snakes is an important sense, it is not as effective as in most other vertebrates, which achieve accommodation – the ability to focus sharply onto the retina – by altering the shape of the

eye's lens using contraction of the muscles of the ciliary body which lies beneath the iris. Most snakes exploit a different mechanism. Their lenses, hard and almost spherical, are difficult to deform so cannot readily be used to focus. So snakes accommodate by contracting the muscles of the iris which increases pressure within the eye and forces the lens backwards and forwards, rather like a camera. In addition, the cells lining the retina are less well organised in snakes than in most other vertebrates; although they have a wide field of vision, which can cover an angle of approximately 130°, snakes do not see detail and seem to ignore immobile objects. So if you ever happen across a snake, the best way to avoid being bothered by it is to keep perfectly still!

Most intriguingly, some snakes can survive quite well even when deprived of sight. This is true for tiger snakes (*Notechis scutatus*) living on Carnac Island off the coast of Western Australia. Many of them have their eyes destroyed by birds defending their nests, which the snakes attack to eat eggs, but the blind tiger snakes seem to live and to reproduce quite well, so the loss of the important sense of vision apparently does little to reduce their viability.

Some sea snakes, like *Aipysurus laevis*, which is commonly found in the waters off the Australian Great Barrier Reef, have photoreceptors or 'eyes' in their tails; the organs are receptive to light but cannot focus it into an image, so they are not real eyes. We do not know why they are present but perhaps it is because during the day, when the snake rests under rocks, the sensation of light might lead it to hide its tail and hence to protect itself from sharks and other predators.

Tasting and smelling

Snakes characteristically extend a flickering forked tongue from their mouths, a highly developed ability which enables them to 'smell' their immediate environment. The tongue captures volatile molecules which are picked up by Jacobson's organ in the roof of the mouth. The Jacobson's organ or vomero-nasal organ works in conjunction with the nostrils and the olfactory part of the brain. It lies below a larger olfactory chamber, and consists of a pair of sacs lined with sensory cells, located in the front of the palate. The sacs open to the roof of the mouth via a pair of tiny ducts and their inner ends connect to separate branches of the olfactory nerve. Different chemicals evoke different nerve responses which are transferred to the brain. This smelling system plays an important role in the life of a snake, notably in identifying an appropriate sexual partner.

Hearing

In humans, the ear comprises three major regions, the external, middle and inner ears; this allows us to perceive frequencies ranging from about 20 Hz to 18,000 Hz, a very low rumble to a high-pitched squeak. Like most reptiles, snakes have no external ear and a poorly developed middle ear, though they do have an internal ear which is rather similar to those of other vertebrates. Recent experiments have indicated that this rudimentary apparatus allows them to detect frequencies between about 100 Hz and 1,000 Hz. They certainly perceive vibrations through their lower jaw when it is in contact with a surface but it is not clear whether this perception is auditory or 'tactile'.

Reproducing

Snakes have a long reproductive apparatus which ends in two hemipenises in males and two hemiclitorises in females. In some cases, snakes copulate for hours and even, in the case of some vipers, for more than two days! Strictly speaking, all snakes hatch from eggs, approximately 70 per cent of

Figure 5.4 Eggs laid by a sea snake *Laticauda semifasciata* collected in South Japan, on Iriomote Island.

which have a solid shell and are laid by females for external incubation; most snakes are therefore said to be *oviparous* or egg-laying. This is the case of the *Laticauda*. Thus, in New Caledonia a few years ago, I visited a small island in the lagoon of Nouméa. In a rocky cavity, I found a nest of peaceful specimens of *Laticauda laticaudata*, bodies banded blue and black, with a clutch of six white eggs. It was a wonderful scene. Also a nice suprise was the discovery of four eggs in the bottom of a bag where a *Laticauda semifasciata*, collected in a Japanese island, was kept for a while (Figure 5.4).

But the females of some species retain the eggs inside their bodies to produce fully developed live young like ourselves (Figure 5.5). They are called *ovoviviparous*. This strategy offers considerable advantages for snakes which live in cold countries, but is also seen in some snakes native to warmer climes, like boas. It affords clear advantages for marine snakes whose offspring are born underwater. Immediately after birth, the young swim to the surface where they breathe for the first time.

Skin sloughing

The sloughing of skin is not unique to snakes. We, too, constantly lose dead epidermal cells. The main difference is that snakes get rid of their cells synchronously, losing their entire skin in one go, from the tip of the snout to the

Figure 5.5 Parturition in *Crotalus durissus*. The fully developed live young is wrapped in a transparent membrane (D. Heuclin).

Figure 5.6 Sloughing of the viper *Bitis nasicornis* (D. Heuclin).

tip of the tail, including even the caps covering the eyes (Figure 5.6). One can predict when sloughing will occur because one or two weeks before it starts the snake usually stops eating and its eyes assume a whitish hue. A snake shedding its skin usually rubs its nose against a rough surface until a piece of skin is lost and then jigs up and down against something solid like a stone, branch or the ground until it emerges from its sheath.

Diet and eating

Depending on what they eat, snakes can be roughly categorised as generalists or specialists. As an illustration, consider some examples from the *Colubroidea* superfamily, a large group of snakes that includes all venomous species. Overall, their diet is highly heterogeneous with a wide variety of invertebrates such as molluscs (including slugs and sometimes snails), crustaceans (crabs and crayfishes) and various types of insects. They may also feed on vertebrates, among them amphibians, reptiles (lizards and snakes), small mammals and birds. Some snakes may eat various types of eggs, an intriguing situation as we shall shortly see because the egg can be at times larger than the head of the snake (see pp. 38–39).

Viperidae (vipers) predominantly feed on small mammals like rodents, but this is not their only source of food. The African puff adder (*Bitis arietans*) occasionally eats ground-living birds and toads. The African big viper *Bitis gabonica* may even take small antelopes. Vipers may also prey upon insects and in the summertime the European viper (*Vipera ursinii*) feeds on grass-hoppers that are plentiful in some mountains. Rather more exotic is a unique North American *crotalid*, a mocassin of the genus *Agkistrodon* which likes decaying carcasses.

Cobras, mambas, kraits, coral snakes and other terrestrial elapids (see Chapter 6) tend to be generalised feeders, eating small vertebrates such as lizards, toads, frogs, small mammals and slow-moving fish. This is particu-larly so with snakes of the genus *Naja*, including the African *Naja nigricollis*. They may raid fowl runs for eggs, sometimes killing distressed fowl in the process. Some snakes are real 'gourmets'! In season, the Australian *Notechis* preys upon chicks of mutton-birds. A number of terrestrial elapids, including coral snakes, cobras and Australian elapids eat other snakes (Figure 5.7a). Certainly, the Asian King Cobra, *Ophiophagus* (which means 'snake-eater') *hannah*, is one of the greatest snake eaters (Figure 5.7b). Arboreal mambas may eat birds and some tree-dwelling mammals; it is sometimes claimed that these mambas fall from a tree right onto their prey – amazing if true but it is unclear that it really happens.

Some sea snakes, like *Aipysurus laevis* or *Lapemis hardwickii*, dine on various types of fish, cuttlefish, squid and crabs, whereas other sea snakes are food specialists. Thus, sea snakes of the genera *Laticauda* mostly feed on eels while those of the genera *Emydocephalus* are exclusively egg-eaters.

Not unexpectedly for snakes living underground, the *Atractaspididae* prey on a variety of burrowing reptiles, frogs and small rodents, as well as lizards.

Colubridae live in various habitats throughout the world and display a great heterogeneity of feeding habits – for example, arboreal back-fanged African snakes like the dangerous boomslang (*Dispholidus*) – while the slender *Thelotornis* chooses birds, their nestlings and even their eggs, which are usually swallowed whole. These snakes also feed heavily on chameleons and other tree lizards and may take bats and other small mammals. In Aus-tralia, the tree snakes *Boiga* feed on lizards, birds and their eggs, and small mammals, whereas the aquatic *Cerberus* eats small crabs and fish. In Europe, the venomous 'Couleuvre de Montpellier' (*Malpolon monspessulanus*) feeds heavily on vipers.

Snakes do not chew prey. Swallowing is a slow process achieved by a striking combination of successive movements of various bones of the upper and lower jaws. Usually, the mandible opens downwards, the skull rises and the bones of the roof of the mouth push forward. The teeth of the palate enter the prey and a series of relative movements of the independent mandible and palate begin, giving the impression that it is the head which moves forward relative to the swallowed prey.

(a)

(b)

Figure 5.7 Cannibalism in snakes. (a) The non-venomous California Kingsnake, *Lampropeltis getulus*, eating a venomous rattlesnake, *Crotalus cerastes*. Note that the Kingsnake is partially immunised against the haemorrhagic action of some venoms (D. Heuclin). (b) The King cobra, *Ophiophagus hannah*, known to feed on other snakes (R. Stöcklin).

This is impressively illustrated by *Dasypeltis inornata*, a small member of the Colubridae (we will discuss these a little later, pp. 62–64), which preys exclusively upon birds' eggs (Figure 5.8a). An egg with a diameter as much as two and a half times that of the snake's head is swallowed whole (Figure 5.8b–d). In the oesophagus the egg shell is crushed by the *hypapophyses*, spike-like projections of the first few vertebrae which extend down into the throat. Then the contents of the egg are pushed down towards the stomach whilst the shell is regurgitated (Figure 5.8e). The system is nicely controlled

(a)

(b)

(c)

(d)

(e)

Figure 5.8 The non-venomous snake *Dasypeltis inornata*, ready to eat an egg (a). The snake can open its mouth wide (b, c), swallow the egg whole (d), break the egg's shell, swallow the white and yolk and regurgitate the broken shell (e) (D. Heuclin).

so that no shell fragment is swallowed and no egg contents are regurgitated. Regurgitation is frequently observed with most snakes that are disturbed after feeding.

Perceiving heat

Some snakes possess something like radar. Those of the subfamily *Crotalinae* have an infrared-sensitive detector in the shape of the pit organ located between the eye and the nostril, rather nearer the latter (Figure 5.9). This thermosensitive organ is remarkably efficient and responds to temperature changes of 0.003°C in 0.1 second, allowing snakes to localise 'warm-blooded' prey even in complete darkness.

Messaging

Snakes also play 'music'. The rattlesnakes *Crotalus* and *Sistrurus* have a tail which ends in a series of loosely attached keratinised segments; when shaken, they produce buzzing or rattling sounds (Figure 5.10).

Snakes in and under water

Sea snakes enjoy a number of specific adaptations. Imagine the problems of an air-breathing animal remaining underwater for 30 minutes or more, during which time prey must be hunted and eaten. How can seawater be prevented from entering the nostrils and penetrating the lungs? Sea snakes

Figure 5.9 A rattlesnake, *Crotalus viridis*, with its pit organ, a hole located close to the eye, with which the snake can rapidly detect very small temperature changes. A convenient 'radar-like' tool to detect prey (D. Heuclin).

Figure 5.10 A rattlesnake, *Crotalus viridis*, displaying its 'musical' rattle (D. Heuclin)

have specific flaps ('valvular nostrils') attached to the rear border of the nostril, which bend inward when open. When they are closed, the nostrils become watertight and airtight: water cannot enter or air cannot leave and the snake can swim under water before resurfacing to breathe. It is a remarkable sight when a snake does surface, expels air with a characteristic sound, breathes and dives again. That sea snakes can stay underwater for such long periods is explained by their slow metabolism, as well their single but well-developed two-chamber hollow lung which provides adequate oxygen storage.

Some sea snake species are excellent divers; within five minutes or so they can reach a depth of 100 m and remain there for quite a while. After a prolonged dive at depth, the blood has high concentration of dissolved nitrogen. If a human diver surfaces too quickly, the nitrogen is released and forms bubbles in the bloodstream, often causing the sometimes fatal condition known as 'the bends'. It is unclear whether snakes also suffer from the bends.

Like sea fish, sea snakes are not bothered by salt water. In contrast to the halophilic ('salt-loving') archeabacteria, animal cells cannot live in too salty an environment. A whole organism like a sea snake can live in the sea because a variety of processes keep down the concentration of salt in their bodies. First, their skin is highly resistant to the loss of fresh water and to the ingress of salt. Second, they expel the brine by a specialised gland under the tongue which concentrates and then excretes it into the sea through the tongue canal. The internal concentration of salt is thus maintained below that of seawater. Other reptiles living in salt water (sea turtles and crocodiles, for example), also possess salt glands, although not under their tongues: in sea turtles it is in the orbit of the eye, thereby giving the strange impression that turtles are always crying.

Venom apparatus

Some snakes possess two venom glands in the upper jaw, beneath the scales and behind the eye. These highly differentiated glands may have evolved from salivary glands. Connected by appropriate ducts to fangs located in upper part of the mouth, the glands are surrounded by muscles which compress them when venom is to be delivered. The fangs are teeth which have evolved to this function over time; they are sharp and can readily penetrate tissues; the venom is ejected through small longitudinal holes or grooves.

Four distinct regions can be distinguished in the venom gland of a viper (Figure 5.11). The main gland occupies the largest part, nearly two-thirds of

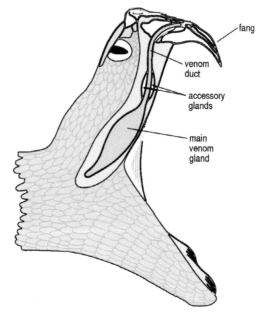

Figure 5.11 A schematic representation of the venom apparatus of a viper. Adapted from Ernst and Zug in *Snakes in Question*, published by the Smithsonian Institution, 1996.

the whole venom gland. It is connected to a small accessory gland by a *primary duct* and that gland, in turn, is linked by a *secondary duct* to the sheath of the *fang*. In the elapids, the venom gland is usually divided into two parts, the *rear main gland* and the *anterior mucous gland* with a secretory duct. The venom is produced in the main gland which contains tubules converging to the centre of the gland and opening into a small cavity or lumen where the venom is stored.

In general, cobras, mambas, kraits and related species have short, hollow fangs on their maxillary bones, with inlets at the bases and outlets near their tips. The venom is ducted to the inlet and forced through the hole at the tip of the fang. We have seen that some snakes, like *Naja nigricollis*, can spit their venom as a defensive behaviour (Figure 5.12). They rapidly contract certain muscles and force venom through small apertures in the front of the fangs, thus squirting it out at high speed as small droplets up to 2.5 m. The fangs of the spitters are a little peculiar with a round outlet of reduced size in the front of the fang, above the tip (Figure 5.13). When pressure is applied to the gland, the venom is very rapidly squirted through the outlet; it may be sprayed a few metres with a fair degree of accuracy. Although the venom has almost no effect on contact with skin, it can cause painful irritation if it reaches the eyes and rapid rinsing is essential to avoid damage to the cornea.

Figure 5.12 The African cobra, *Naja nigricollis*, spitting its venom (D. Heuclin). The spat venom can be seen on the right of the figure, 'floating' in the air.

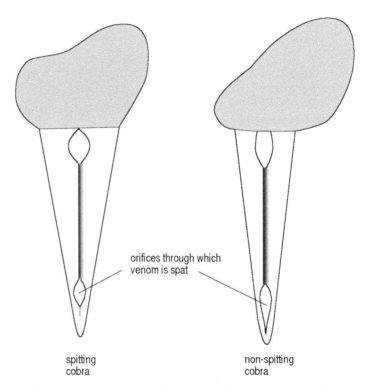

orifices through which
venom is spat

spitting
cobra

non-spitting
cobra

Figure 5.13 Front view of the fangs of a spitting (left) and non-spitting (right) cobra. Only the orifice through which the venom is spat really differs. It is more rounded and shorter in spitters. Adapted from Ernst and Zug in *Snakes in Question*, published by the Smithsonian Institution, 1996.

A snake bite delivering no venom is called a 'dry bite'. Even when venom is injected, not all of the snake's supply is necessarily delivered: a viper may inject only 10 per cent of its total supply while a cobra might introduce as much as 60 per cent and an Australian elapid up to 80 per cent. Once the venom has been used up, the gland has to be refilled; this takes several days and, surprisingly, the rates of synthesis in a specimen vary for the different venom components. The rate of synthetic activity is controlled by the amount of venom actually present in the gland. For the moment, the detailed molecular processes controlling the triggering of venom synthesis remain unclear.

Where snakes live
Snakes dwell in meadows, temperate, tropical or rain forests, savannahs, deserts, high mountains, deep valleys and in water, including oceans and

coral reefs. Some live in trees, others on the ground or indeed underground. Even in regions where they are comparatively rare, there is a fair chance that one day you will come across a snake, most likely in the countryside and perhaps nowadays even in urban centres, given the number of people who keep them as pets.

Some snakes congregate both on land and in the sea. We have all seen movies in which the hero falls into a pit of writhing snakes. Remember how Indiana Jones, still a junior in *The Last Crusade*, fell through the roof of a train into a box filled with snakes which crawled all over him, even inside his shirt? Such films use snakes that are totally harmless; the non-venomous garter snake is one of them, a North American colubrid of the genus *Thamnophis* which has beautifully coloured longitudinal stripes. When they stop hibernating, these snakes crawl around in their hundreds.

Snakes get eaten, too

Snakes have predators, including birds, mammals, crocodiles, alligators, and even snakes themselves, so it is hardly surprising to find that many of them have a remarkable ability to camouflage themselves with sophisticated coloured patterns (Figure 5.14).

Equally extraordinary is their ability to adopt different postures, one of

Figure 5.14 A beautifully coloured non-venomous snake, *Lampropeltis triangulum* (D. Heuclin)

Figure 5.15 An Asian cobra, *Naja phillipinensis*, spreading its hood (D. Heuclin)

Figure 5.16 An albino Asian cobra, *Naja kaouthia*, spreading its hood (R. Stöcklin).

Figure 5.17 A stuffed Asian cobra, as can be found in curio and tourist shops throughout the tropics.

the most impressive being that of the cobras, which rear up and spread their hoods (Figure 5.15 and Figure 5.16).

However, the decline in snake populations is largely due to human interference. Some people consider snakes as delicacies and they are also used in some folk medicines and, intriguingly, as aphrodisiacs. In Europe and in Asia, bottled snakes preserved in alcohol are not a rare sight in tourist shops. Smoked snakes are also prized in some places and I was once lucky enough to join in the collection of sea snakes off a small Japanese island near Okinawa. One night, my Japanese friends and I waited for the tide to fill one of the many rocky galleries surrounding the island. With the tide came the sea snakes, *Laticauda semifasciata*. These gentle, peaceful and beautifully banded animals rest and lay eggs on land, and are easy to catch; they rarely bite. We collected a number and took them to the local ruler, a sort of a queen called Noro, who allowed us to milk the snakes to collect venom. We took a few specimens back to the laboratory while the remaining snakes were smoked and subsequently exported around the world, a practice commercially important for these people.

Stuffed snakes are all too common in curio and tourist shops throughout the tropics (Figure 5.17), sometimes mounted in a fighting pose with a stuffed mongoose. One would think that taxidermy would destroy the venom but the juice extracted from venom glands of a stuffed cobra is still toxic to mice. Snake toxins are extremely resistant to destruction so be careful the next time you are tempted to buy an apparently innocent souvenir – it may not always be harmless! Indeed, the commercial use of some snakes is now so intensive that they are actually endangered and worldwide regulation is urgently needed if we are to avoid their complete extinction.

6

Classification of snakes

Taxonomy is the theory and practice of classification. It can be applied to all sciences and, indeed, to other fields, but many people think of it mainly as applied to plants and animals. Nature comprises such enormous variety that we need ways of sorting things out and establishing relationships. But with millions upon millions of species, this becomes very complicated, revisions are frequent, and it is easy to lose one's way. Here I shall just use broad brush strokes to paint a picture of snake taxonomy. Let me start by telling you a short story.

Imagine a biochemist who receives a vial containing venom from an African black-necked spitting cobra dressed up with its fancy name *Naja nigricollis*. The biochemist isolates the major toxic principle from the venom and establishes its chemical structure. Not unreasonably he calls it 'toxin from *Naja nigricollis*'. It may be that subsequent taxonomical studies show that the original designation of the snake was too simple, or too complex, and that its name needs to be changed. Under the new classification it might become *Naja mossambica pallida*, and perhaps still later, after another classificatory revision, *Naja pallida*. Should the name of the toxin follow these changes? A biochemist interested in the chemistry of toxins should be aware that the name of their snake had changed. However, he or she is likely to want to keep the toxin name unchanged because the chemical structure of the protein is the same whatever the name attached to it.

Taxonomists may take another view and lively exchanges can ensue! The best way of dealing with this situation might be to give a defined compound isolated from venom a particular or specific name, one not dependent on the specific name of the source. For example, the names *erabutoxins a, b* and *c* were given to the three toxins isolated from venom of the sea snakes *Laticauda semifasciata*. Erabu is the name of the island that harbours the sea snakes from which the toxins were first isolated. Once a name is fixed, anyone interested will be able to follow it through the scientific literature.

Phylogeny of animals
All creatures that share the six following attributes belong to the animal kingdom:

1   they are multicellular;
2   they obtain carbon and energy by feeding upon other organisms, some-
    times upon each other, and on organic wastes;
3   they need oxygen;
4   they reproduce sexually or asexually;
5   their lifecycles include a period of embryonic development; and
6   most are motile, at least during part of their lifecycles.

Having identified the common characteristics of animals, establishment of a
refined phylogeny depends on the identification of major differences, allowing us to group animals first into broad *phyla* and then into smaller *classes,
families, genera* and so on.

   One obvious criterion to use for categorising animals is the presence of
a backbone: if it has one it is a *vertebrate*, if not, an *invertebrate*. However,
this is a very broad classification, dividing the whole variety of animals into
only two groups and we have to go further. Modern animal classification is
usually based on the following five basic characteristics.

Body symmetry
Some animals, like jellyfish, have a so-called *radial symmetry* with their body
parts organised regularly around a central axis like the spokes of a bicycle
wheel. Others, like humans, have a *bilateral symmetry*, with right and left
halves that are mirror images, usually with a head and a posterior part, a
back and a ventral surface.

Cephalisation
In association with bilateral symmetry, pairs of sensory structures and nerve
cells are usually concentrated in the front end, the head.

Type of gut
Food is digested in a gut. Some animals, like sea anemones, have sac-like
guts with one opening only for both taking food and expelling waste. Other
animals have tube-like guts with an opening at each end (a mouth and an
anus).

Type of body cavity
Many animals have a clear bilateral symmetry, with a body cavity separating
the gut from the body wall. One type of cavity is called a *coelom*. We
humans have a coelom divided into thoracic and abdominal cavities by the
diaphragm, a sheet of muscle.

Segmentation
'Segmented' animals like earthworms possess a series of body units often, though not always, similar to one another. The body of other animals, like insects, can be organised into three regions (head, thorax and abdomen) that greatly differ from one another.

These criteria are the basis for dividing animals into phyla shown in most zoology textbooks. However, more recent data based on comparative analyses of gene nucleotide sequences suggest an even more refined phylogeny. Such a tree is shown in Figure 6.1 in which venom-producing animals, including snakes, are highlighted. Snakes belong to the phylum of chordates

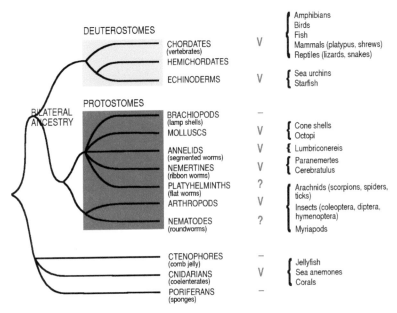

Figure 6.1 A phylogenic tree of the Animal Kingdom based on the analyses of some of their genes. Each phylum that includes venomous species is indicated by a red V. Animals whose bodies have bilateral symmetry are divided into protostomes and deuterostomes. This division is based first on the way the mouth of the animals is generated during embryogenesis but also on a number of other characteristics. Thus, protostomes have ventral nerve cords whereas deuterostomes often have a dorsal neural tube. Names in parenthesis are just rough indications to provide the reader with a better appreciation of what sort of animals the formal name of a phylum corresponds to. The phylogenic tree is based on various works, including one by Adoutte *et al.*, 1999, published in *Trends in Genetics*, Vol. 15, p. 104.

– animals with a segmented bony or cartilaginous spinal column. Within this broad phylum, snakes belong to the *group* or *subphylum* of vertebrates which includes some 40,000 living species of fish, amphibians, reptiles, birds and mammals; very many others are now extinct. But this is just the beginning of the classification.

The subsidiary categories then commonly used by taxonomists are orders, suborders, infraorders, families, subfamilies, genera and species, and sometimes also subspecies. The concept of species is not straightforward. In 1940, Ernst Mayr proposed a definition that is still broadly accepted: a natural group of living creatures whose individuals can effectively or potentially interbreed freely and are reproductively isolated from other similar groups of living organisms. For example, horses can *mate* with donkeys but cannot *breed* with them because the offspring of such matings (mules) are sterile; horses and donkeys are therefore separate species. All categories of human beings can interbreed with one another so humans are all one species. Closely related species form a genus and, in turn, closely related genera form a family, etc. In common usage the genus name begins with a capital letter and the species a small letter; thus *Naja nigricollis* and *Homo sapiens*.

Dentition, an old but still valuable tool for the identification of snakes

Anyone can easily differentiate between two animals such as a sea snake and a rattlesnake. However, simple observation may not always be sufficient to identify to which group of snakes the animal belongs, or even whether it is indeed a snake (for instance, the 'blind-snakes' we mentioned in Chapter 1). The snakes themselves form a complex group with many subdivisions, each defined on the basis of agreed characteristics and properties.

In the mid-nineteenth century, it was suggested that snakes be classified into four groups according to their dentition. This classification is simple and sometimes still used by specialists. What can be seen in the mouths of different snakes?

Vipers and atractaspids (see below, pp. 61–62) possess the sophisticated and impressive *solenoglyphous* (tubular-fanged) dentition (Figure 6.2). In addition to a number of small teeth, both at the back of the upper jaw and along the lower jaw, these snakes have two tubular venom-conducting fangs, the longest being as much as 5 cm in length. These fangs are the only teeth of a small maxillary bone located at the very front of the mouth; this bone is articulated and, when the mouth opens, the fangs can be erected nearly perpendicularly to the skull bones, a posture highly appreciated by photographers. Great care is required with close-ups since an attacking viper can erect its fangs in one-twentieth of a second and project its head at a speed of 36 kph an hour (22 mph)!

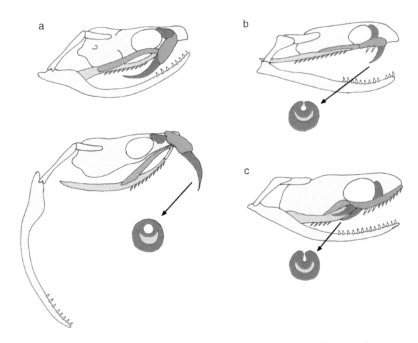

Figure 6.2 Dentition of venomous snakes. (a) Vipers have a solenoglyphous dentition. The longest and tubular teeth, called the fangs (coloured red) are located at the very front of the mouth on the maxillary bone which is hinged (see top and bottom) and hence can be erected almost perpendicularly to the skull bones. (b) Cobras and other marine and terrestrial elapids have a proteroglyphous dentition, with grooved fangs on the front edge of almost immobile maxillary bones. (c) Venomous colubrids have an opistoglyphous dentition with grooved fangs at the rear of the maxillary bone. A transversal section is shown for the different fangs. The maxillary bone, which is uniquely articulated in solenoglyphous snakes, is shown in blue. The other homologous bones in the different snakes are coloured similarly: the pterygoid in yellow and the ectopterygoid in purple.

The elapids (see below, pp. 58–60), like cobras and mambas, have *proteroglyphous* (front-fanged) dentition with a pair of fixed fangs at the front of the maxillary bone followed by a few small additional teeth. These fangs are deeply grooved and often enclose a canal for venom ejection (Figure 5.13).

The venomous colubrids (see below, pp. 62–64), like the African boomslang, have an *opistoglyphous* (rear-fanged) dentition with grooved fangs located at the rear of the jaw and connected by a duct to the venom-producing *Duvernoy's gland*. The maxillary bone of opistoglyphous snakes often includes several small teeth in front of the fangs which therefore seem appropriate for injecting venom after the prey has been caught.

A rat snake has an *aglyphous* (fangless) dentition, with no specialised venom-conducting teeth. Aglyphous snakes include all those considered non-venomous. Although they have no obvious fangs, some of these colubrid snakes nevertheless possess a venom-producing Duvernoy's gland.

The distinction between opistoglyphous and aglyphous snakes is not always clear. Some snakes belonging to the same phylogenetic group may possess either of these dentitions. Moreover, while Duvernoy's gland is a possible precursor of other venom glands, it does seem to be more closely related to the venom glands of proteroglyphous snakes than of solenoglyphous snakes.

Modern methods of classification

A sensible classification system needs objectively defined criteria for species, genus and so on. Current classification is based on the use of external and/or internal discriminative features called 'keys'. In the case of snakes, various morphological features can be used. Consider two examples.

A characteristic of the African genus *Naja* is the presence of one or two upper scales (called *upper labials*) bordering the lips and entering the orbit. Among the identifiers of the sea snake genus *Laticauda* is the number of scales along an imaginary line around the middle of the body but excluding the enlarged ventral scales: there are 21 in *L. colubrina* and 19 in *L. laticaudata*. If the snake is venomous, the number of maxillary solid teeth behind or in front of the fangs is also an important and clear identifying feature.

Other sources of information include *in vitro* fertility studies, determination of chromosomal complementarities and comparisons of amino acid sequences of related snake proteins of similar function: how closely related, for instance, are the structures of haemoglobin, the blood pigment protein? Recent studies have shown that venomous snakes can be classified on the basis of the amino acid sequences of their venom proteins such as potent neurotoxins: the molecular biologists' ambition to sequence every genome in sight will perhaps have the most profound effect in establishing the ancestral relationships between species and even individuals. No systematic sequence analysis of a whole reptile genome has yet been undertaken, though many independent studies on individual genes are under way. But it can only be a matter of time: gene sequencing is becoming easier, cheaper and more efficient. In the meantime, existing methods of snake classification remain fairly reliable, allowing us to place snakes pretty accurately in the living world.

Three infraorders for all snakes

Snakes belong to the phylum of *Chordata*, the subphylum of *Vertebrata*, the class of *Reptilia*, the order of *Squamata* and the suborder of *Serpentes* (or *Ophidia*). In 1987, after much work and a lot of hesitation, the suborder of the Ophidians of more than 3,000 species was divided into three infraorders (see Table A5 in Appendix). One is *Cholophidia*, which comprises all fossil snakes not included in the other two infraorders.

A second infraorder is *Scolecophidia* (the name comes from the Greek *scolex* 'worm' and *ophis* 'snake') which consists of three families of small (15–100 cm long), worm-like, burrowing and harmless snakes. They live in tropical and temperate regions, digging small tunnels with their heads. The eyes are small spots under the scales and the mouth is under the head. Living hidden from sunlight, their eyesight is limited and some of them are apparently sightless, thus explaining why in English they are called 'blind snakes'. They are found from Mexico to Iran as well as in Australia, Pacific islands, Europe and Southern USA.

The third infraorder is *Alethinophidia* (from the Greek *Alethinos* 'genuine' and *ophis* 'snake'). The *Henophidia* is one of the groups that form this infraorder. It comprises the world's largest snakes, such as pythons and boas. Palaeontological studies indicate that these animals, some of which may have reached 15 m in length, were the dominant snakes on earth from 65–23.5 million years ago, after which they began to decline in number. Among the surviving species, the pythons, boas and anacondas, are still large creatures and may grow to 10 m long. Living in various aquatic, terrestrial, arboreal and burrowing environments, they are well known for their ability to coil around their prey, which die from suffocation or heart failure: the *Boa constrictor* from South America is one such example. Whilst the *Henophidia* were declining, the *Colubroidea* superfamily, which includes all venomous snakes, was expanding and diversifying.

In spite of all the work which has been done, the classification of snakes is far from being settled and, indeed, will become much clearer with comparative analyses of snakes genes, including those coding for their toxins.

A single superfamily for all venomous snakes

All the venomous snakes belong in the *Colubroidea* superfamily which probably appeared not more than 53–35 million years ago. It is a huge group, comprising about 2,300 species possessing a number of specific characters. The heads are clearly distinct, the mouth is usually large, the maxillary bones are longer than the skull, the mandibles can open widely and they have no vestige of posterior limbs. Distributed worldwide, they may be terrestrial, arboreal, burrowing and aquatic. Approximately 600 species are venomous and the bites of about 250 can cause severe problems in humans.

The superfamily of *Colubroidea* includes four families of living snakes: the *Elapidae, Viperidae, Atractaspididae* and *Colubridae* (which should not be mistaken for the Colubroidea superfamily: a confusing near-identity of spelling). Each family has a number of venomous snakes (see classification of snakes in Table A5 of the Appendix).

The Elapidae are proteroglyphous snakes with two relatively short-grooved or hollow fixed fangs at the front of their maxillary bones. With 60–70 genera and nearly 280–300 species, classification is not an easy task. Recent studies suggested the existence of two subfamilies. The *Elapinae* include the terrestrial elapids of Africa, Asia and the Americas. The *Hydrophiinae*, include the terrestrial elapids of Australia and New Guinea, as well as all sea snakes and sea kraits. To keep matters simple we will arrange them into two artificial sets, the terrestrial *elapids* and the marine *elapids*.

The terrestrial elapids are a diverse group of snakes (Figures 6.3a–c) with more than 55 genera and 230 species, widely distributed in tropical areas in a range of environments: arboreal like mambas, terrestrial like most cobras and burrowing like coral snakes. They have always stimulated human imagination and were long used in fantasy stories and even in advertisements, so much so that some of them are quite familiar to everyone. Who has never seen a photograph of a spectacular African or Asian cobra (genus *Naja*), which may attain a length of 2–3 m? Particularly striking is the cobra's aroused posture: head raised and hood open, suggestive of nobility and fascination (see Picture 5.15). Popular literature is also full of the 'frightening' black or green African mambas (*Dendroaspis*), up to 4 m long, and the equally long 'frightful' Australian taipans (*Oxyuranus*), not forgetting the Australian tiger snakes (genus

(a)

(b)

(c)

Figure 6.3 A. Some elapid snakes. (a) The Australian tiger snake, *Notechis scutatus* (D. Heuclin). (b) The Asian krait, *Bungarus fasciatus* (D. Heuclin). (c) The Australian brownsnake, *Pseudonaja textilis* (J. White).

Notechis) (Figure 6.3a), Asian kraits (*Bungarus*) (Figure 6.3b) and the smaller but highly coloured coral snakes from Asia (*Maticora* and *Calliophis*) or America (*Micrurus* and *Micruroides*). Among the most imposing is the impressive King Cobra (*Ophiophagus hannah*), the largest venomous snake, which feeds on other snakes and whose classification among the elapids remains surprisingly unclear. This snake can reach an extraordinary 5–6 m in length (Figure 5.7b). I could not fail to be moved when I saw one such specimen in the snake farm of the Thai Red Cross Hospital in Bangkok, particularly when one of the farm keepers entered its cage as it raised its head nearly a metre and a half off the ground, disturbingly close to the keeper's head! But then stories about King Cobras abound.

The sea snakes should not be mistaken for sea serpents; nor are they mythical creatures to be confused with fish. They breathe air like you and I, albeit with lungs that have special adaptive features. There are 15 genera of 'true' sea snakes and one genus of sea kraits (*Laticauda*) (Figure 6.4). Their pattern of distribution is interesting as they are totally absent from the Atlantic and Mediterranean but abundant in the warm waters off Australia and Malaysia. One of them, *Pelamis platurus*, has the largest distribution of all: from the waters off the west coast of the Americas to the coastal waters of east Africa and northern Australasia. New species of sea snakes are discovered from time to time and there is an abiding interest in how they cope with the difficulties of marine living.

Figure 6.4 A sea snake, *Laticauda schistorhynchus*, swimming in the shallow waters of the coral reef around Niue, a small and remote island in the Pacific.

The *Viperidae* constitutes quite a well-known family of snakes with more than 30 genera and 230 species. You may have heard the old saying taught to European children living in the countryside: only vipers have a V-shaped mark on their head. Though rather a simplistic view, the posterior part of the viper head does indeed tend to be larger. They have the typical solenoglyphous venom apparatus with their fangs being located on a movable bone. Widespread throughout Europe, Africa, the Middle East, India, Asia (Figure 6.5) and

Figure 6.5 An Asian viper, *Trimeresurus albolabris* (D. Heuclin).

the Americas, they are not found in Australia, Papua New Guinea or nearby regions. They live in quite diverse environments: a number in tropical regions, others at high altitude (close to 5,000 m [16,400 feet]), and one, the common European viper, within the Arctic circle.

The family of *Viperidae* also includes the famous pitvipers, a name that may call to mind rattlesnakes and, while this is true, not all pitvipers have a 'musical' tail. The subfamily *Crotalinae* includes 19 genera and about 150 species and is absent from Africa and Australia but rather common in various parts of Eastern Europe, Asia, India and the Americas. The large *Crotalinae*, *Lachesis muta*, the bushmaster which lives in forested countries from Nicaragua to South America, is thought to be related to rattlesnakes.

The *Atractaspididae*, also called *atractaspidids*, form a third family of venomous snakes which, however, is much less known, and its sub-classification is still unclear. Some authors claim that the family includes eight genera and about 60 species, while others assert that it has only one genus and at least 15 species, *Atractaspis*, also called the 'burrowing asps', 'stiletto snakes' or 'molde vipers'. The *Atractaspis* are highly venomous and possess a solenoglyphous venom apparatus and for this reason have long been incorrectly categorised together with the vipers. These dark snakes are found in Central and South Africa, the Middle East, Sinai, Jordan and the southern part of Israel where they usually live in long narrow, underground tunnels (Figure 6.6).

How can they bite their prey in narrow tunnels? In fact, their venom apparatus is beautifully adapted to their habitat. The snake strikes its prey by jabbing sideways and backwards with fangs that protrude from its almost closed mouth. They are tricky to handle and are certainly not safe to be held by the neck behind the head, as one can do with most other snakes, because they can still bite and their venom is quite potent. These snakes are probably responsible for many accidents among amateur herpetologists.

The *Colubridae*, also called colubrids, is a large group of close to 290 genera and almost 1,700 species. These snakes are found in nearly all parts of the world and encompass both venomous and non-venomous species. It is the major snake group virtually everywhere except in Australia where the *Elapidae* predominate. At least two-thirds of them possess Duvernoy's glands of appreciable size which produce a toxic secretion; these glands are different from the venom glands of elapids and viperids.

A number of colubrids are typical rear-fanged opistoglyphous snakes. Most are not dangerous to humans, but there are a few exceptions, like the famous arboreal African boomslang (*Dispholidus typus*) (Figure 6.7) whose unusually small maxillary bone means that its back fangs are not far from the front of the mouth. It can open its jaws as wide as 170 degrees and easily

Figure 6.6 An African atractaspid, *Atractaspis microlepidota andersoni* (D. Heuclin).

Figure 6.7 An African colubrid snake, the boomslang *Dispholidus typus* (D. Heuclin)

seize a leg or arm, and its venom is quite potent, making the boomslang highly dangerous.

Other colubrids have aglyphous dentition with no grooved fangs. In some a duct emerges from the Duvernoy's gland on the external face of the maxillary bone, opening near simple solid and ungrooved teeth. This is not a very efficient venom apparatus and the toxic secretion sometimes enters the prey's body as it is chewed during swallowing.

Colubrid snakes live in a wide range of habitats and feed on a variety of prey, including worms, snails, crabs, small vertebrates like rodents, fish, birds and even other snakes. They range in size from about 20 cm for the worm-eating snakes (*Carphophis*) to 3.5 m for the Asian rodent-eating snake (*Zaocys*).

7

Discovery of snake venoms

Love is an odd the word with which to start a chapter on snake poisons but the word 'venom' presumably has the same root as 'Venus'. It comes from the Latin *venenum* meaning a decoction from a magic plant, a charm or filter, and was rapidly associated with the notion of poison. *Venenum* itself might originate from *venes-num* for 'love filter'.

In one sense, the bite of a snake is, in principle, no different from that of dogs, cats and crocodiles which may subsequently become infected by viruses, bacteria or other microorganisms. But in practice there is a difference: when a venomous snake bites it can deliver a poisonous fluid. What is this venom, where does it come from, is it the same in all snakes, are all snake venoms dangerous for humans and how do they exert their fatal effects? These are some of the obvious questions which have been debated over the centuries. Answers have emerged slowly, with much difficulty and sometimes only after considerable dispute. Let us start at the beginning of the seventeenth century.

The first discovery of venom
About three centuries ago, it was believed vipers were poisonous because they could emit 'irritated spirits' when biting. These spirits were supposed to be so 'cold' that blood coagulated in the veins, so fatally blocking the circulation. Views that we now find very strange were common at that time because many basic phenomena were quite mysterious. The simple notion of gas was just emerging, pioneered by the Flemish chemist and physician Jan Baptista Van Helmont. And to explain most common physical observations, the German chemist Georg Ernst Stahl described the *phlogiston theory*, an unbelievable monument of strangeness which persisted for a long time. Stahl thought that, during combustion, something was expelled from the burning material but also that an 'expeller' was needed. He called this principle *das verbrennliche Wesen* ('the combustible principle'), to which his followers gave the Greek name *phlogiston*, the flame. It came as a bit of a shock when oxygen, hydrogen and nitrogen were later discovered but the phlogiston

theory was so embedded in people's minds that for a while nitrogen was called *phlogistic air*. Complex living organisms, and their associated phenomena such as snake bites, were even more difficult to explain, so much so that, quite naturally, metaphysical or religious notions were attached to them. So why not to snake spirits?

Nicolas Lemery, a French authority on pharmacy during the second part of the seventeenth century, thought that many great men, including Van Helmont, based their belief that snakes emitted spirits on so much experience that they needed no further studies to support their opinions. It therefore took a lot of courage to contradict this established dogma and the bold step was eventually taken around 1670 by Francesco Redi, an Italian medical doctor born in 1626 in Arrezo in Tuscany. Redi was a great scientist who struggled against metaphysical explanations for a variety of physical phenomena. In particular, he disproved the theory of spontaneous generation which held that worms and maggots emerge as if by magic from rotting meat. Redi not only believed this idea to be false but, like the true experimentalist that he was, set out to disprove it. He simply placed pieces of meat in two vials and covered one of them: maggots arose only from the uncovered rotting meat. When the flies were prevented from landing on the meat, no maggots grew. According to Redi, 'the maggots were all generated by insemi-nation and the putrefied matter in which they are found has no other office than that of serving as a place, or a suitable nest, where animals deposit their eggs at the breeding season and in which they also find nourishment.' Not all scientists, though, were totally convinced by Redi's conclusion and humanity had to wait for the famous experiment by Pasteur, two centuries later, for spontaneous generation to be definitely disproved.

Redi also found unacceptable the idea that the hazard of venomous snakes was due to their emitted spirits. To disprove this notion, he again designed simple and brilliant experiments, collecting the yellow juice from a viper's fangs (Figure 7.1) and applying it to wounds deliberately inflicted on various animals. All the treated animals died and Redi quite naturally con-cluded that the juice was the venom. And he was right: his work marked the real start of scientific studies on snake venoms.

Having shown that a specific 'liquor' was responsible for the lethal power of vipers, Redi next asked two simple, related questions: 'how does the venom act?' and 'where does it come from?' He thought about various possible modes of action, including coagulation of the blood in the heart or veins or both. So puzzling was the problem that he claimed the question was 'far above his own strengths and that probably he would never know the correct answer'. He was convinced that the reservoir of the venom was located in the membrane that covers the fangs. One English scientist, Richard

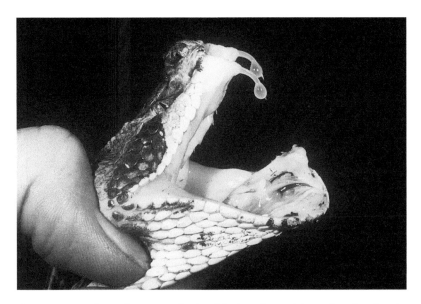

Figure 7.1 Two drops of venom at the tip of the long fangs of the rattlesnake, *Crotalus horridus* (D. Heuclin)

Mead, was of a different opinion and it took some time before the solution was found. Although the general outline of a viper's venom apparatus was known in the eighteenth century, it was not until 1924 that the detailed mechanism was understood.

Fighting with so many accepted dogmas, Redi remained nevertheless deeply attached to the basic beauties of life. This exceptional scientist was also a celebrated poet who wrote many verses on wine, a liquor much more popular than snake venom.

> Ma di quel che si puretto
> si vendemmia in Artimino
> vo'trincarne più d'une tino.

Amazingly, today many Italians know him more for his poetic talent than for his scientific career.

Having identified the active poison, the next question was, how does it work? About a century after Redi's original experiments, another Italian, Felix Fontana, and John Hunter, an American, took up the problem. In his *Traité sur le vénin de la Vipere* (*Treatise on viper venom*) published in 1781, Fontana described the results of more than 6,000 experiments with 3,000 vipers. He observed that, upon administration of large amounts of viper venom into the jugular vein of rabbits, the animals screamed, manifested violent convulsions and died within a few minutes. The animals' blood was black and coagulated in all the vessels as well as in the heart; viper venom clearly acted on blood. A few years earlier, however, other scientists had noticed no coagulation when dogs or cats were bitten by the same type of viper. It was not until the end of the nineteenth century that Césaire-Auguste Phisalix, a French medical doctor, demonstrated that viper venom can cause coagulation when injected at high doses and anticoagulation at low doses.

Not all snake venoms act the same way. As we shall see later (pp. 100–102), about 100 years ago several researchers established that paralysis, the major action of venoms from cobras, kraits and sea snakes, is comparable to that seen in animals poisoned with curare. Some snake venoms, therefore, contain extremely potent curare-like substances. These substances have played an essential role in molecular pharmacology, leading in the early 1970s to the isolation of the target of curare, an essential protein which regulates muscle contraction. Many other individual components of snake venoms have since been isolated and shown to act on a large range of essential properties of the neuromuscular system. Snake venoms thus exert diverse actions on various physiological systems.

Something else intrigued Fontana: are vipers sensitive to their own venom? He concluded they were not and extended the lack of poisonous effect of viper venom to various colubrids and some molluscs, including snails. Over the last century, a number of scientists have disagreed with those results but recent experiments demonstrated that the circulating blood of snakes from various families contains proteins which neutralise the major toxic activities of snake venoms. The chemical structure of these neutralising components is now known. Snake blood thus contains proteins, decoys almost, which prevent snake venom from reaching its target. This explains Fontana's observation that vipers were unaffected when injected with their own venom.

An outstanding question remained regarding the nature of viper's venom. It was particularly challenging in Fontana's time because modern chemistry was only just emerging (Antoine-Laurent Lavoisier of France, often regarded as its founder, was just 38 years old when Fontana wrote his book) and the nature of the components of living organisms was particularly mysterious to scientists of the period. Doubtless frustrated after all his experiments, Fontana concluded very honestly that 'The venom of a viper, like many other compounds, is a substance that includes several principles that we do not yet know anything about'. He also wrote: 'How far are we from piercing this mystery? How many difficult and unknown pathways will it be necessary to explore before we shed some light on this complex and obscure matter that is venom?' Through these words Fontana identified two major lines of research on venoms: what is their nature and how can their components be separated? These questions have kept researchers busy for nearly two centuries.

Snake venom is made of proteins

The nature of venom was determined during the nineteenth century. After much debate, it was agreed that venoms are *protein*-like substances and not alkaloids or 'cobric acid', as originally suggested by some chemists. The Swedish Baron Jöns Jakob Berzelius, a brilliant chemist who discovered several chemical elements and introduced a system of chemical symbols, suggested the word *protein* in a letter sent to the German scientist Gerardus Mulder, another chemist.

The trouble was that nobody knew anything substantive about protein structure. The substances underlying anatomical or physiological chemistry, i.e. the stuff of living organisms, were not simple to define and, in the middle of the nineteenth century, their chemistry was obscure. A late nineteenth-century dictionary defines albumins (a type of protein) as 'the soluble principles present in egg white, blood serum, lymph, milk, urine, plant extracts, etc., which crystallises when heated'. It also defines albuminoids as 'the principles of living organisms which contain nitrogen. These included all albumins, muscular fibrin, the alcohol-soluble part of gluten, all soluble ferments and so on.'

In 1841, Mulder proposed that albuminoids are composed of a common 'radical' that interacts with various amounts of sulphur, phosphorus, oxygen or salt. As suggested by Berzelius, Mulder named this essential radical *protein* which in Greek means 'primary'. The belief in an invariable and common radical in all albuminoids was, in fact, incorrect and was subsequently abandoned, although we now know, of course, that there are important common features in the structures of all proteins (see Appendix). Nevertheless, the name 'protein' survived and is now used to define the class of polymers made up of amino acids, just like albumin, the principal component of egg white.

There is another proposal for the origin of the word protein, one which is imaginative but unfortunately wrong. The argument goes that the word might have come from Proteus, the son of Poseidon the god of the waters, earthquakes and horses who could change his shape at will. Some proteins do indeed modify their shapes, a feature that is often a prerequisite for expression of their biological activity, but the name nevertheless came from Berzelius and Mulder.

During the nineteenth and twentieth centuries, many scientists worldwide tried to separate venom into its protein components. Proteins are not

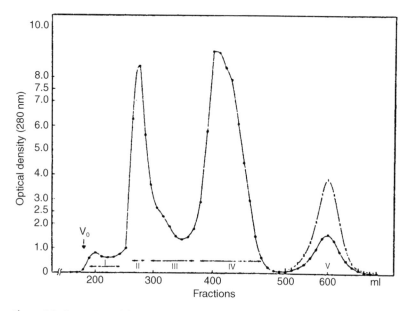

Figure 7.2 Separation of the components of the venom purchased at the Pasteur Institute under the name, *Naja nigricollis*, by chromatography on a column of
gel called Sephadex® G75. One gram of venom was loaded on top of a gel-
filled column 1 m long. The proteins were eluted with a solution of ammonium acetate, pH 7.3. Proteins of molecular weight above about 75,000 are
eluted first from the column (peak I). The components present in peaks II,
III, IV and V are progressively smaller. Fraction IV contains the components
that are lethal to mice. However, this separation remains rather crude, as
each fraction still contains many components. This is illustrated in Figure
7.3 where the proteins present in peak IV have been submitted to ion
exchange chromatography of greater resolution.

easy to handle and conventional methods successfully applied in organic
chemistry were unsuitable for venoms. Despite the difficulty of the situation,
a number of excellent advances were made, one of them by Lucien Bona-
parte (a brother of Napoleon) who, in 1843, used alcohol and ether to isolate
a toxic fraction from viper venom. The properties of this fraction, called
viperine, were comparable to those of 'digestive ferments' (the universal biological catalysts now called *enzymes*). Around 1860, S. Weir-Mitchell in the
USA isolated *crotaline*, another toxic fraction, by boiling Crotalidae venom
and treating it with alcohol. Then, the Australian physiologist Charles Martin
separated the venom of an Australian snake into two fractions, one of which
caused haemorrhages while the other induced respiratory failure, demon-

strating that the venom possessed physically distinct properties. However, until the middle of the twentieth century, and despite great efforts, chemists failed to isolate really pure proteins that could account for the toxic properties of snake venoms. New methods were clearly needed and, in 1956, Elbert Peterson and his group introduced *ion exchange chromatography* (see box and Figure 7.3), which can be used to separate proteins in aqueous solution according to their electrostatic charges. Almost at the same time, it became possible to separate proteins by molecular size. In 1959, Jerker Porath and Per Flodin in Sweden introduced the cross-linked dextran as a chromatographic material, so creating the technique of *gel filtration* (see box and Figure 7.2). It was soon after this that the first venom toxins were separated and identified.

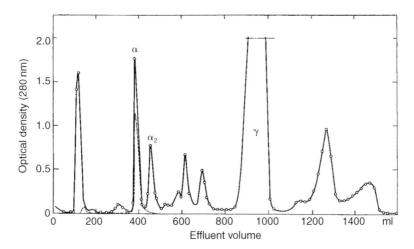

Figure 7.3 Ion exchange chromatography of the proteins present in fraction IV from Figure 7.2. A 25 cm column filled with a resin called Biorex (which contains negatively charged groups) is equilibrated in 0.05 M ammonium acetate pH 7.3 and subjected to a linear gradient of the same salt, up to 1.4 M. Proteins with much negative charges are not retained on the resin and are immediately eluted in the void volume (the first peak on the left). Increasingly basic proteins are then eluted in order. Thus, the curaremimetic toxins, α and $\alpha 2$, are eluted before γ, the cardiotoxin, in agreement with isoelectric focusing findings (see Figure 11.13). Note that the cardiotoxin is more abundant than the curaremimetic toxins, although they are much less potent, at least in mice.

Chromatography

There are many types of chromatography but all are based on the idea of a liquid flowing past a stationary solid which itself holds liquid in a stationary form. Chemicals introduced with the flowing solvent then spend part of their time in the moving liquid and part in the stationary one. The two liquids are not identical and the proportion of time which each chemical spends in each phase depends on the chemistry of the substance and that of the two liquids. The longer the chemical spends in the stationary phase, the slower it will travel through the column or tube.

An early separation in 1906 by Mikhail Tswett in Russia (and the one which gave the procedure its name – *chromatography*, meaning 'coloured writing') was that of leaf pigments on columns of crushed chalk. The chalk was wetted with water and the pigments, dissolved in a small amount of an organic liquid which could mix to some extent with water, were placed on top of the chalk and allowed to soak in. Then more of the organic liquid was slowly and continuously poured through the column.

Pigments which could dissolve in the water held stationary by the chalk were slowed down in their passage through the column compared with those which had to remain entirely in the organic liquid: the more completely pigments dissolved in water, the more time they spent in the chalk and the slower they moved. Because the pigments were coloured, it was possible to see bands progressing down the column and, indeed, each pigment could be separately collected as it came out.

Other procedures use different means of separation but the name 'chromatography' has stuck for all of them. In 'ion exchange chromatography', the material making up the separatory columns are electrically charged, so attracting oppositely charged compounds in the mixture and slowing their passage through the column. 'Gel filtration' uses yet another approach, this time to separate mainly molecules of different sizes. In this case the column material comprises hollow carbohydrate beads with pores of graded sizes. Small molecules can enter the pores and, because it takes them time to find their way out, their passage through the column is delayed. Molecules too large to enter the pores flow unencumbered through the column and emerge more rapidly at the bottom.

8

Snake venom potency

Philipus Theophrastus Aureolus Bombastus von Hohenheim, a Swiss physician born in 1493, was and still is known as Paracelsus, a Greco-Roman
translation of 'Hohenheim' which means 'next to heaven'. For Paracelsus,
everything was a poison. He wrote in his famous *Third Defence*:

> 'What is there that is not poison?
> All things are poison and nothing [is]
> without poison. Solely the dose
> determines that a thing is not a poison'

All substances are dangerous, even those we like most, but only if we get too
much of them. In other words, the amount administered – the dose – of any
substance, pure or impure, is a critical parameter governing its toxic or non-
toxic action. The concept of dose is central to understanding toxicity and is
usually expressed as the amount of the substance administered to an organism per unit of bodyweight.

Toxicity is the dose of the substance causing a toxic reaction, sometimes
defined as the minimal dose at which an individual person or animal mani-
fests the first signs or symptoms of the toxic action. However, responses are
affected by weight and a variety of other factors and vary a lot in different
individuals.

To measure toxicity we start by selecting a sensitive species. You will
remember that rabbits are sensitive to viper venom while vipers themselves
are much less so. White mice weighing 20 g are often used as test animals.
Next, the substance needs to be administered to a sufficient number of individuals to compensate for variability of individual sensitivity. Third, a route
of administration needs to be established: oral, intravenous, intra-muscular or
intra-cranial. Fontana showed more than 200 years ago that viper venom is
ineffective when swallowed but is highly active when injected into the
jugular vein of a rabbit. Other variables include the time to onset of the toxic
action, the sex and age of the individuals and the ambient temperature. The
method that was and still is sometimes used to determine toxicity consists of

estimating the injected dose per unit weight (per gram or kilogram) causing death in 50 per cent of the test animals within a given period. This is the *median lethal dose*, or LD_{50}.

Unfortunately, these methods require the use of living animals. They were and still are absolutely necessary in the first steps of the study of toxic substances. However, when the physiological target has been identified, laboratory methods of response measurement can be developed, thus avoiding the use of living creatures.

Another obvious requirement for estimating venom potency is purity. Some venoms and toxins are now available commercially but scientists may wish to collect others themselves. This takes special expertise but is not especially difficult, except with some particularly troublesome snakes. The snake is first immobilised, for instance by pinning its neck with a forked stick and firmly grabbing its head with the other hand. This may not be advisable with some snakes, like the *Atractaspididae*, whose fangs can bite you through their closed mouths. Great care is always needed: milking a snake can be quite a risky business and it is certainly a good idea to seek expert advice the first time you try it.

Figure 8.1 Collection in Fiji of venom (almost 0.2 ml) from *Laticauda colubrina*, by inserting the snake fangs separately into a pipette.

If the snake's mouth is large enough, it may be possible to insert a glass vial covered with transparent plastic into which the snake can bite. If the snake is small and has tiny fangs, one can insert each fang separately into a pipette and collect the venom, which is often yellow in colour (Figure 8.1). Sometimes the handler can feel the movement of the snake's muscles when it compresses its venom glands. Sometimes the snake refuses to expel its venom but can be persuaded by applying gentle pressure to the glands which are usually easily located. Milking can be repeated regularly so long as the snakes are given enough time to refill their glands. Excessive milking may result in damaging stress to the animal.

Snake venoms can be viscous and colourless, as for sea snakes, or yellowish, as for other snakes. The volume of venom that can be obtained by milking varies and can be as much as 1.16 ml or as little as 0.08 ml on the first milking after capture.

Venom can be freeze-dried and kept as a powder for years, provided it is protected from light and moisture; samples collected in Vietnam as long ago as 1930 are still quite active. For research purposes it is best not to mix venoms from different individuals of the same species. There may be some variability between venoms and there is always the possibility that the snakes may later be subdivided as different subspecies or even, as different species: we saw earlier that snake taxonomy is subject to regular revisions. One needs finally to bear in mind that venom obtained by milking may not be pure as it is impossible to exclude the possibility that it contains saliva.

Measuring the action of snake venom in a mouse is not necessarily a guide to its effects on humans. Thus, 0.0004 mg of venom from the Australian inland taipan is enough to kill half the animals within 24 hours when administered intravenously to 20 g white mice. Although simple extrapolation suggests that 1.4 mg of this venom would have a 50–50 chance of killing a 70 kg human, humans may be more (or less) sensitive to the venom than mice.

It is acceptable, however, to compare *in mice* the toxic potency of the Taipan venom with that of other venoms, showing that snake venoms vary in potency. For example, to kill 50 per cent of individual 20 g white mice takes 0.002 mg of *Atractaspis engaddensis* venom, 0.01 mg of venom from an Ethiopian spitting cobra, and as much as 0.2 mg of venom of the eastern diamondback rattlesnake (*Crotalus adamenteus*).

Since venom is intended to subdue prey, we may wonder whether dietary variation correlates with variation in the proportion of the different poisonous components present in the venom of a given type of snake. Such a relationship was recently reported in the case of the Southeast Asian viper *Calloselasma rhodostoma*. Venom was collected from 67 adult snakes from 36 places in Vietnam and the different constituents of each sample were separated. The observed variation in venom composition seemed to be linked to geographical variation in diet. However, there is some debate about this observation and it has not yet been proven whether it applies to all snakes.

Can the venom itself represent a danger to the snake if, for example, the snake is wounded and venom enters its blood? An experiment I myself witnessed in 1974 in Japan is revealing. Nearly 1 g of a potent toxin from sea snake (*Laticauda*) venom was injected into a small specimen of the same species weighing about 200 g; that snake received enough toxin to have killed a quarter of a million 20 g mice! The snake nonetheless survived several months after the injection, suggesting that it was protected against its own toxin, probably because the target protein in the prey and the homologous protein in the snake itself differ at least to some degree. Thus, the snake's protein does not interact with its own toxins, a sophisticated relationship between snake, toxin and prey. As we will see later, it is not the only way a snake can protect itself nor does it imply that all snakes are insensitive to all snake venoms. The reality is probably much more complex. For example, venom from snakes that feed on other snakes seems to be highly toxic to the snakes on which they feed but is the snake-eating snake itself sensitive to its own venom? A systematic study of this interesting problem may shed much light on the evolution of snakes and venoms.

9

Clinical aspects of snake poisoning

Humans are generally afraid of snakes because they associate them with the injection of venom, an often well-justified fear. Some 250 species may cause venom poisoning in humans, also termed *envenomation*, or *envenoming*. Although it is difficult to estimate annual mortality accurately, a recent estimate by Jean-Philippe Chippaux was more than five million snake bites worldwide, four million of them in Asia, one million in Africa and more than 350,000 in the Americas. Half of the bites result in various degrees of poisoning and it is estimated that venomous snakes kill 125,000 people every year: 100,000 in Asia, 20,000 in Africa and 5,000 in the Americas. There are fewer than 350 deaths annually due to all snake bites in Europe, the Middle East, Oceania and Australia combined.

If not much venom is injected, the effect will be mild but, if the worst does happen, what will be the effects of poisoning? Rather than go into detailed pathology, we will look just at what a doctor sees in a snake bite victim. The actual patterns vary greatly from one case to another and treatment appropriate to a given victim can only be defined by a doctor after assessment of all the symptoms which, in turn, may also provide useful pointers in identifying the venom components.

Bites by terrestrial elapids

Poisoning by some African cobras, coral snakes, mambas and kraits usually have little or no effect at the site of the bite but they may cause the progressive paralysis which is symptomatic of neurotoxicity. In general, the most highly innervated muscles are the first to suffer: the muscle that elevates the upper eyelid and the ocular and ciliary muscles controlling the lens. Between 20 minutes and several hours after being bitten, the victim may have a real struggle to keep his or her eyelids open, with other common manifestations being vomiting, blurred vision, headache and hypersalivation. The symptoms of paralysis may worsen, with progressive involvement of various muscles, including those of the jaws which, in some cases, become locked. Respiratory distress may occur as a result of paralysis of the diaphragm, the main muscle controlling breathing, although patients may be kept alive and conscious on artificial respiration even if paralysis is complete. Neurotoxic symptoms may resolve spontaneously within a week or can be allayed more rapidly through administration of antivenin or drugs that inactivate *acetylcholinesterase*, an enzyme which naturally destroys the chemical messenger carrying signals from nerves to muscles at the *neuromuscular synapse*.

Most African spitting cobras cause swelling and *necrosis* (tissue death) at the site of the bite rather than paralysis, whilst most Asian cobras usually bring about local necrosis and hardly any neurotoxicity. All of these venoms contain high concentrations of cytotoxic components which affect tissue membranes as well as small amounts of more potent neurotoxins.

Poisoning by Australian elapids gives rise to major and diverse clinical symptoms. Brown snake bites can result in a defibrination syndrome (also called *coagulopathy*). More rarely they may also cause paralysis. The bite of a tiger snake may induce local pain, oedema (swelling) and occasionally local necrosis, followed by convulsions, hypotension and neurotoxic symptoms. Muscle damage and renal failure can also occur. With death adder snake bites, severe neurotoxic changes may begin to occur within an hour. Taipan venom can elicit local swelling, necrosis, convulsions, coagulopathy, severe paralysis, muscle damage and renal failure. Australian snake venoms are rich mixtures of toxic compounds with a range of haemorrhagic, myotoxic, procoagulant, anticoagulant, antiplatelet and neurotoxic activities.

Bites by marine elapids

In spite of all the recreational activity occurring in the warm tropical waters of the Asian Pacific regions, sea snake bites are rather uncommon: the risk of a bather being bitten by a sea snake in Penang Island is unlikely to be greater than one bite in 31 years of bathing. Anxiety is a common early consequence of sea snake bite accompanied by a feeble pulse, rapid shallow breathing and even collapse. But this may not actually be the direct result of venom poisoning and some doctors have recommended the use of a placebo injection to calm patients down.

Sea snake venoms contain toxins acting on both nerve and muscle. A bite by a sea snake is painless, with little or no local swelling, but within an hour or two the victim may experience muscle aches and pains, sometimes muscle spasms, stiffness and tenderness accompanied by headache, thirst, sweating, a feeling of coldness and vomiting; progressive paralysis can subsequently develop. Tonic spasm of the muscles of mastication may also keep the jaw locked in so-called *lockjaw* or *trismus*. Three to six hours after the bite, myoglobin, the form of coloured protein found in muscle fibres, may be present in the urine although this does not always occur.

Bites by vipers

Vipers may cause severe local effects with swelling that can spread rapidly along the whole bitten limb. Necrosis (tissue death) may occur at the site of the wound over the following days, and can be particularly severe after bites by some African vipers (Figure 9.1), Asian pit vipers, lance-headed vipers and rattlesnakes. Although there may be no topical signs after some crotal bites, death might nonetheless occur.

Continuous bleeding from the puncture wound often follows a viper bite. Extensive bleeding commonly also occurs in other parts of the body, including the gums, and is often accompanied by clotting. Severe cases of bites by European vipers may sometimes cause extreme shock, starting no more than 10 minutes after the bite. North American rattlesnake and viper bites often cause hypotension (a drop in blood pressure) in the victim while the eastern diamondback rattlesnake (*Crotalus adamenteus*) usually produces particularly marked swelling and oedema, necrosis and high tissue and blood loss. Haemorrhagic effects (leakage of blood) are also common after bites by Malayan pit vipers. Russell's viper venom may cause fatal renal failure. Occasionally, poisoning by some vipers and rattlesnakes generate fatal neurotoxic symptoms. Viper venoms thus contain various components that perturb haemostasis (control of blood flow), fibrinolysis (clot breakdown) and the neuromuscular system.

Figure 9.1 Necrosis at the site of a wound caused by a bite by an African viper, *Vipera lebetina*. The picture was taken 24 hours after the bite (J.-P. Chippaux).

Bites by atractaspids
As with vipers, burrowing asp poisoning is associated with marked local symptoms, including blistering, swelling, necrosis and pain. A severe bite may induce nausea, vomiting, diarrhoea, respiratory distress, hypertension and marked alteration of the heart beat, sometimes followed by death within one hour. This venom contains potent toxins with strong cardiovascular effects.

Bites by colubrids
The most famous of the *Colubridae* is perhaps the African boomslang. A severe bite by this snake, or by any of a number of other dangerous relatives, may cause local swelling and extensive bleeding from and within the mucous surfaces. Victims sometimes faint, vomit and suffer extreme abdominal pain. Although the venom is quite potent, it can at times act slowly, perhaps taking more than 24 hours to cause serious symptoms, the victim's blood becoming uncoagulable with consequent severe haemorrhage and renal failure. The venom includes toxins acting on blood coagulation and haemorrhagic factors.

Snake venoms are clearly mixtures of a great variety of toxins. Their identification and the determination of their chemical structures, how and on which bodily targets they act, and how they developed through evolution are all of great interest. We will explore these questions later. For the moment we need to know how snake bites can be treated.

Ancient treatments of snake poisoning
One hot August day in the year 30 BC, Octavian, the founder of the imperial Roman government, met Cleopatra. His victory over Egypt was almost complete. All that remained was to convince Cleopatra to join him in Rome and celebrate his triumph. For Cleopatra the only way to escape such dishonour was death, through which she could also join her beloved Mark Anthony. Despite Octavian's best efforts to prevent her from committing suicide, Cleopatra was found lying on her golden bed, dressed in her royal clothes. Just what happened is uncertain but legend has it that she chose to be bitten by the snake of Amon-Ra, which conferred immortality.

In those remote times could she have been saved? In ancient Egypt long before Cleopatra, especially during the second millennium BC, various procedures were already in use to treat snake bites. They probably involved the use of plant extracts and mineral substances and were commonly associated with religious incantations and prayers to convince the snake to withdraw its venom from the bite. Equally common was the prescription of beverages that had contained a papyrus on which was written an appropriate magic formula. How effective these treatments were is not recorded! Thereafter, as civilisations developed, the first pharmacopoeias and physical interventions devoted to the treatment of snake bites progressively emerged.

Celsus, a Roman medical author of the first century AD, gave a remarkable description of his recommendations for physical interventions to be employed in the case of snake bite. Once bitten by a venomous snake, he recommended application of a tourniquet to the affected limb, without excessive tightening so as to avoid numbing the limb. Superficial incisions were then to be made at the site of the bite to stimulate bleeding. This could also be accentuated by sucking blood with a suction-cup or even with the mouth. The victim was placed in a warm room, the bitten limb being orientated downward. Medications could be applied to the wound and, if necessary, the limb was amputated.

Ancient medications

A medication that dominated practice for many centuries was *theriac* (sometimes known as *treacle*), an antidote composed of various substances of mineral, plant or animal origin. Its name derives from the Greek *theriake*, from *therion*, a wild animal. Among the famous doctors whose names were associated with this medication are Mithridates who, around the first century BC, designed a complex mixture of more than 50 plants and minerals that he called *mithridatium*; and Andromachus the Elder, Nero's personal physician, who made *great theriac*, which also comprised of many ingredients including a large dose of opium (which certainly helped!) and certain viper extracts. Some 16 centuries later, doctors continued to prescribe a similar type of mixture to treat various diseases, including poisoning by snakes. The addition of viper extracts was based on the assumption that some parts of the bodies of venomous animals would contain the antidote. A number of scientists of the seventeenth and eighteenth centuries criticised this type of remedy but it had become so deeply engrained in medical usage that many years passed before it vanished. It was finally removed from the official French and German lists of medical treatments only in 1884 and 1930, respectively.

But theriac was not the only remedy for snake bites. An impressive list of substances was proposed as efficacious antidotes. Most popular was *mummy*, a substance initially extracted from Egyptian mummies and then from any dead body. Various solid substances were also recognised as important antidotes: one of them, introduced into Europe from Arabic medicine, was the *bezoar*, a stone found in the digestive apparatus of some ruminants. A number of sixteenth century doctors like Ambroise Paré denied its efficacy but it nevertheless remained highly popular for another 200 years. Believe it or not, there is another solid substance, the *snake stone* or *black stone*, which is still prepared by certain religious groups as an antidote in some parts of the world. This is a burnt cow bone which, when applied to a scarified bite, sticks until it has absorbed all the venom. Once full of venom, it loses its efficacy but can, it is believed, be endlessly regenerated in milk. I am sure you agree that the idea is appealing but whether it is actually effective remains to be seen.

Yesteryear's popular traditions live on
There are many other unproven and, in fact, potentially dangerous 'traditional' treatments. It is surprising that more or less irrational remedies continue to be used in treating snake bites. One recent example is electric shock: a few years ago, a respected scientific journal published an article advocating the application of an electric shock at the site of a snake bite as an appropriate treatment. Although their efficacy remains to be demonstrated, electrical devices to treat snake bites (and poisonous insect stings) are commercially available. The manufacturer recommends the application at the site of the bite of a 25,000 volt electric shock for 2 seconds, with 10 seconds between shocks. Moreover, the assumption that suction removes venom has engendered aspirating devices currently sold in pharmacies but, once again, their efficacy is doubtful.

Why do people still believe in these sorts of things? There may be at least two major reasons. First, when facing a difficult situation like a snake bite, there is an urgent need for the victim to be reassured and this is easier to do if the victim believes that a simple and quick intervention will resolve the difficulty. Moreover, it is not impossible that the resulting confidence triggers some placebo-type beneficial effects. Second, it is clear that not all snake bites involve venom injection; it is likely that about half of them are *not* venomous. Even if injection does occur, the amount of venom introduced may be small. Thus, in these situations, application of any traditional treatment will give the effect of efficacy, further fuelling belief in its appropriateness. The downside is that these methods may instil overconfidence and, in the event of real poisoning, may result in unsuitable treatment by the victim or his/her unskilled well-wishers.

New approaches to the treatment of snake bites should not be discouraged, but should be subject to rigorous scientific evaluation and offer at least some advantages over existing methods.

The proper treatment of snake bites
There are two steps in the treatment of bites:

1   how to behave just after a bite in the absence of professional medical
    care; and

2   clinical management by professionals.

It is beyond the scope of this book to describe what should be done for each specific type of bite but it is worth outlining the general approaches accepted by most specialised doctors. A detailed description may be found in *Clinical Toxicology of Animal Venoms and Poisons* published by J. Meier and J. White in 1995.

 World Health Organisation specialists have proposed general guidelines for the treatment of all types of animal poisoning.

1   Prevent panic. The victim may be frightened and, from personal experience, I can assure the reader that this is indeed the case. Never forget to reassure the victim that not all bites contain venom and that the medical care which will soon follow is always highly effective.
2   Immobilise the bitten limb with a long crepe bandage to avoid accelerated spread of venom. Then carefully transport the victim to hospital.
3   *Never* tamper with the bite wound or the victim: no suction, no incisions, no tourniquet or constricting band, no electric shock, no injection of any medicines or chemicals of any kind, no ice packs on the wound, no snake bite kit, no alcohol, no food and no drugs (no aspirin or sedatives). If there is likely to be a long delay before medical treatment is possible, give the victim water to avoid dehydration.
4   If possible, bring the dead snake to the hospital for identification but be careful when handling it.
5   Seek medical advice as quickly as possible.
6   If necessary provide air ventilation and external cardiac massage. The victim may faint.

As soon as the victim arrives at the hospital, professionals take over. Their first priority is to maintain life. Snake bite emergencies may require basic resuscitation: airway establishment and control, and restoration of circulation. In other cases a patient may look quite well but nevertheless develop the signs and symptoms of snake bite poisoning several hours later. Careful observation is therefore required to identify particular problems such as paralysis and renal failure (blood in urine). Specific treatment may then be prescribed.

 When available, antivenin should be injected. If not, a number of specific agents may be used. These include an inhibitor of acetylcholinesterase when curaremimetic toxins from terrestrial or marine elapids cause paralysis. If the patient's diaphragm is paralysed, ventilation may be vital, although usually for not more than a few hours.

 However, this is a very simplified description. Do not imagine that treatment is always straightforward.

Antivenins: a revolution in neutralising snake poisons
The end of the nineteenth century was a really exciting period for scientists interested in venoms and toxins, and in developing protective methods. Let us have a quick look at the remarkable observations made with toxins from bacteria. In 1888, Emile Roux and Alexandre Yersin at the Pasteur Institute in France demonstrated that a soluble principle, a *toxin*, could be isolated from cultures of diphtheria bacteria. This toxin caused all the symptoms of the disease: it was not the bacteria per se which were directly responsible for the disease but the toxin made and secreted by them. In 1890, Emil von Behring in Germany, the first Nobel Prize winner in medicine, together with his colleague Shibasaburo Kitasato, observed that animals immunised with diphtheria toxin produced something in their blood which could neutralise or destroy the toxin, thus preventing the disease. These protected animals were *vaccinated*, a phenomenon already known from smallpox, a disease which could be forestalled by infecting people with a mild relative of the disease called *cowpox*. A little earlier, in 1887, Henri Sewall, an American scientist, had reported that pigeons treated for several months with low but progressively increasing doses of venom from a particular type of rattlesnake became resistant to a venom dose that would kill an untreated pigeon. The same effect was later observed with other animals, including dogs and rodents, in which resistance to viper venom could be produced by injections of progressively increased sub-lethal doses of venom. The principle of vaccination against venoms and toxins was firmly established.

What exactly happens in vaccination? Consider a simple experiment. Over a long period of time we inject a rabbit with progressively increasing doses of venom, then collect the blood and then the serum, a clear yellowish fluid, by separating the blood into its solid and liquid components. In a test tube, we mix a small amount of this rabbit serum with a lethal dose of venom and inject the mixture into an ordinary, untreated rabbit. This animal is lucky; it does not die since it is protected from the toxic effects of the venom. Another rabbit injected with the same dose of venom mixed with the same amount of serum from an ordinary, untreated rabbit inevitably dies. The serum of a rabbit injected with progressively increasing doses of venom therefore contains a protective factor.

Even more interestingly, a similar protection can also be triggered by injecting venom which has lost its toxic character; detoxification can be achieved in various ways, including treatment by heat or chemical reagents. Treated like that, the venom loses its harmful properties but not its capacity for eliciting the protective factor.

That protective factor can be transferred to an untreated rabbit, as was clearly demonstrated in 1894 by Albert Calmette and simultaneously by

Gabriel Bertrand and Césaire-Auguste Phisalix, three French scientists of the time. A new treatment called *serotherapy* was born, in which a serum from a protected animal is administered to an unprotected one. This general principle is still relevant today and, in fact, is almost the only specific treatment against snake bite poisoning.

It was soon recognised that the effectiveness of protective sera depends on various factors, including the origin and quantity of venom injected by a snake to the victim, the way in which the antivenin is introduced to the patient, the time lapse after envenomation, the site of the bite and the type of venom action to be neutralised. It was also realised that a protective serum could contain a number of undesirable proteins, like albumins, which might generate allergic-type reactions. Antivenins have been greatly improved over the last 100 years, especially through the identification of the molecular principles which afford the protection. Today's antivenins are quite safe and highly efficient.

Antivenin: the result of a complex immune response

It is intriguing that progressive injection of venom slowly but surely triggers production of antivenin which essentially neutralises the injected venom and no others. A serum prepared against a viper venom, for instance, fails to neutralise a cobra venom and vice versa. The factors underlying this very specific protection process are all proteins called *antibodies*, exquisitely tailored to bind to venom components and, in particular, to toxins. Once they are wrapped in a 'layer' of antibodies, the toxins become harmless.

How are antitoxin antibodies produced? The underlying series of complex molecular and cellular events has been elucidated in recent decades. A detailed representation of the process is schematically depicted in Figure 9.2. One of the secrets of the process is that the toxin has 'the capacity to select' the cells that produce the antitoxin antibodies that ultimately neutralise its action.

Huge numbers of cells, called *B-cells* ('B' for bone marrow), already exist in our bodies, pre-programmed to produce and secrete a large diversity of antibodies. At its surface, each B-cell exposes a single type of antibody: the antibody that it is capable of producing. A toxin can recognise and stick only to the sub-population of B-cells that display the antibodies that are exactly complementary to it. However, two major problems remain:

---

1   at this stage the selected B-cells are not yet able to produce the antibodies;
2   the number of selected cells is small so they can only produce a limited amount of antibody.

The selected B-cells therefore need to be transformed into antibody-producing cells and to expand in number. This is achieved in a complex and beautiful way. Under the control of the antigen itself, the antigen-specific B-cells begin to exchange signals with others called *T-cells* and *antigen-presenting cells*. As a result of a precise cellular 'conversation', the B-cells multiply and produce antibodies directed specifically against the toxin circulating in the blood. This is how we are immunised, or protected if you prefer, against toxin-associated diseases, like diphtheria or tetanus.

Each discrete region of the toxin that can select a given species of antibodies is called an *epitope* and the complementary region by which the antibody binds to the epitope is called a *paratope*. A toxin can carry several different epitopes on its surface and can therefore select different species of antibody-producing B-cells. Indeed, a serum directed against a toxin contains a large mixture of antibodies which bind to the various toxin epitopes. Some antibodies recognise overlapping epitopes while others bind to topographically unrelated epitopes. The whole procedure is very complex but highly effective.

There are antibodies against most if not all protein components, including the non-lethal ones, in snake venoms together with a heterogeneous population of antibodies against each component. In principle, only antibodies directed against the harmful components of the venom are expected to be protective and, within this population, only those that neutralise the harmful characteristics of toxins may be useful. An interesting issue to be addressed in future research is the possibility of favouring the selection and transformation of the appropriate B-cells, those that produce the neutralising antibodies.

Prospects in the treatment of snake bites
Poisoning due to snake bites can now be properly managed thanks both to the experience of medical doctors and to the availability of appropriate antivenins. Yet, in some countries, their management remains a major medical problem. New protective methods need to be considered, particu-

larly to improve the quality of antivenins, lower their cost and rapidly diagnose the nature of the venom.

With appropriate antibodies, it is in principle simple to establish the presence or absence of venom after a snake bite. A blood sample is mixed with an antibody which binds only to a given venom; if the complex is formed, the venom is there. There are procedures permitting the detection and quantification of such complexes and, although this approach is not always as simple as it may appear from this description, it does give the attending doctor an immediate indication of whether a given toxin is present, even if no obvious symptom has yet been observed. This *immunodiagnosis* helps the doctor decide on the most appropriate course of action and to determine the type of antivenin, if any, which should be administered. This approach is currently used in Australia and some other countries.

Vaccinations against venoms (that is, treatment of susceptible patients in advance of the possibility of a snake bite so as to palliate its effects should it happen – just like prior vaccination for diphtheria or smallpox) have been attempted in some regions, in particular in the southern part of Japan where a viper called the *habu* constitutes a medical problem. Detoxified venom was used to vaccinate 43,446 volunteers. The toxoid was effective in preventing local lesions caused by venom but the actual efficacy of the vaccine was not clearly demonstrated.

It is perhaps not surprising that vaccination has not been extended to other parts of the world because there are problems to be overcome:

1   snake venoms act quickly, implying that the protective antibodies need to be always present in sufficient quantities in the blood of a potential victim;
2   there are so many different venoms in a given region that it is difficult to envisage vaccination against all of them;
3   preparation of detoxified venoms, the vaccinating agents and the vaccination procedures are costly; unless new, simplified and powerful procedures are invented, vaccination against snake bite is unlikely to be universally adopted as a preventive measure; and
4   the method has not been totally convincing because the vaccine was only moderately tolerated by the patients and its efficacy was not very high.

More than two centuries ago, Fontana reported that 'the venom of the viper is not poisonous for its species'. He also observed that other animals, including other snake species, turtles, and various molluscs, are resistant to viper venom. Since then, it has been found that the mongoose (Figure 9.3) hedgehog and opossum have resistance against snake bites, but only recently have

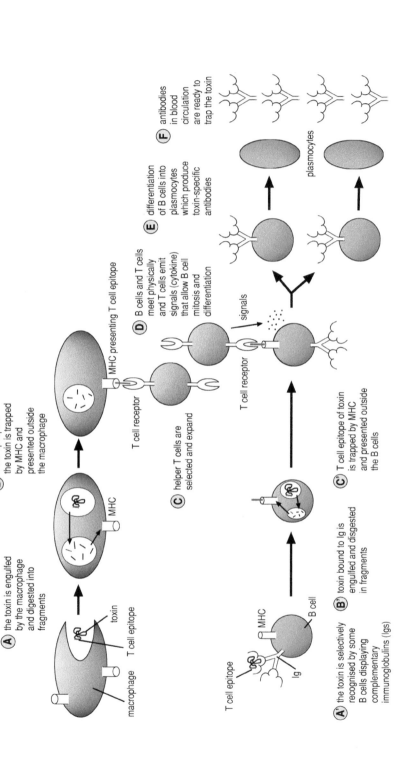

A the toxin is engulfed by the macrophage and digested into fragments

B T cell epitope of the toxin is trapped by MHC and presented outside the macrophage

macrophage   T cell epitope

toxin

MHC

MHC presenting T cell epitope

T cell receptor

C helper T cells are selected and expand

T cell receptor

D B cells and T cells meet physically and T cells emit signals (cytokine) that allow B cell mitosis and differentiation

signals

E differentiation of B cells into plasmocytes which produce toxin-specific antibodies

plasmocytes

F antibodies in blood circulation are ready to trap the toxin

T cell epitope

MHC

B cell

Ig

A' the toxin is selectively recognised by some B cells displaying complementary immunoglobulins (Igs)

B' toxin bound to Ig is engulfed and disgested in fragments

C' T cell epitope of toxin is trapped by MHC and presented outside the B cells

the components conferring this resistance been isolated from plasma. It is not yet known whether any of these compounds could be developed as protective agents against snake bites in humans.

A number of studies have shown that a variety of natural compounds might also be potential neutralising agents of snake venoms. This research is often based on traditional medicines developed in various countries, especially in Africa, India and China. For example, when snake venoms are mixed with stem bark juices from plants like *Schumanniophyton magnificum*, their potency in mice clearly decreases. A small protein of about 60 amino acids, whose amino acid composition resembles that of snake cardiotoxins, has been isolated from that same plant and shown to neutralise the activity of cobra venom. Various flowering plants are used in the treatment of snake bites and a number of compounds extracted from them neutralise venom activity. It is unclear whether the compounds work only when poisoning has developed or if they have to be taken before being bitten. The mechanisms of inactivation remain unknown and finding out might well be a valuable line of exploration.

Heparin has also been recommended to treat snake bites. It is a polymer of sugars present in almost all tissues and especially in the liver. Its use as a potent anticoagulant in snake bite treatment is highly controversial despite a number of impressive successes. In France some years ago there was a well-publicised case of a young herpetologist who opened a bag containing a

---

Figure 9.2 Cellular and molecular events that lead to the production of antibodies against a toxin. The toxin is first 'swallowed' by two categories of cells. Many various cells, like macrophages, directly engulf the toxin *non specifically* (A). A small number of B-cells trap *specifically* the toxin with their membrane immunoglobulins (Ig) that offer a site that is complementary to the toxin architecture (A). The toxin-Ig complexes are then engulfed by these B-cells (B). Once the toxin is internalised, both categories of cells digest it into fragments (B, C). Among the different toxin fragments that are generated, some, called T-cell epitopes, bind to molecules of major histocompatibility (MHC) that are displayed at the surface of both types of cells (B, C). Toxin-specific epitopes are thus 'presented' by numerous 'non-specific' presenting cells to the CD4$^+$ helper T-cells that possess a receptor specific to the T-cell epitope (C). The contact allows CD4$^+$ helper T-cells to expand and hence to increase their probability to meet a toxin-specific B-cell which presents a toxin-specific epitope (C). When a physical contact occurs between the helper T-cells and these B-cells, helper T-cells secrete cytokines (D), which allows their mitosis, proliferation and differentiation (E). These toxin-specific B-cells are thus transformed into cells, called plasmocytes, which produce the antibodies that are specific to the toxin (F). These toxin-specific antibodies then accumulate in blood circulation.

Figure 9.3 A mongoose fighting with a cobra. Mongooses are one of a group of mammals that have acquired some resistance to snake venoms (D. Heuclin).

dangerous venomous snake and was bitten on the face. A photograph of her swollen face was quite widely publicised; she recovered fully after treatment with heparin.

Despite many attempts to find effective treatments against snake bites, serotherapy combined with appropriate symptomatic management remains the therapeutic method of choice. While it is true that a number of improvements have been made in the preparation of antivenins, the basic principle remains quite similar to the one that was introduced by Calmette more than a century ago. Do we really need to discover more efficient means? In the developed world, the problem of snake bites is not pressing but they do remain a major health problem in many tropical countries. We need to keep searching for efficient and simple therapies bearing in mind, nevertheless, that we are not likely to find a panacea against all snake bites because of the great diversity of venom constituents and actions.

10

Non-toxic venom components

When I entered the world of venoms I was most intrigued by a simple observation. Apparently, and in contrast to what might have been expected, snake venoms contained two categories of components. Some are obviously highly lethal to various animals, a mouse for example, whereas others are clearly harmless. While we can understand intuitively that lethal components will help snakes to subdue their prey, what could be the role, if any, of the other constituents?

A snake venom gland is a remarkable factory that produces and dissolves an incredible number of components. Solubilised matter accounts for

Figure 10.1 *Trimeresurus flavomaculatus.* Venomous snakes, such as this one, produce potent toxins but suprisingly their venom also include a number of non-toxic components (D. Heuclin).

almost half of the weight of a snake venom, compared with no more than 1 per cent in the case of our saliva, which is nonetheless considered a rather rich liquid. Venom is therefore a highly concentrated 'juice'. Usually, more than 90 per cent of the solid matter is protein, the rest being inorganic components and small organic compounds.

If cobra venom is submitted to appropriate gel filtration (see box 'Chromatography', p. 72), at least three major fractions can usually be separated (Figure 7.2). One is composed of large proteins, predominantly enzymes (fraction II in Figure 7.2). The second (fraction IV) comprises small proteins, 50–150 amino acids long, while the third and last (fraction V) includes very small compounds. Let us examine the enzyme fraction.

Non-lethal components

At least 26 enzymes have been found in snake venoms. All, of course, are proteins, usually between 150 and 1,500 amino acids long, and many of them do not kill mice within 24 hours. The presence of these non-lethal enzymes led many scientists to propose that snake venoms might contribute to the digestion of the prey, a scenario which fits nicely with the view that venom glands may have evolved from salivary glands. What do these enzymes do?

Connective tissues form the supporting and connecting structures of animal bodies. Their ground substance is rich in large negatively charged polymers, called *proteoglycans*. Proteoglycans consist of repeats of disaccharide units (sugar units) such as *hyaluronate*, linked to proteins. Virtually all venoms possess an enzyme called *hyaluronidase* which can destroy hyaluronate (Figure 10.2); this means that such venoms assist the breakdown of connective tissues in the prey, aiding the snake's digestion and facilitating dispersion of other enzymes and toxins. Many venoms also possess other enzymes acting on sugar-based compounds.

Lipids (fats) constitute a major component of cell membranes and it is striking that nearly all snake venoms include potent *phospholipases* A_2 which might also contribute to the degradation of *phopholipids* (fats which also contain phosphorus atoms) of cell membranes (Figure 10.3).

Nearly all snake venoms include nuclease enzymes which destroy DNA and RNA, the genetic memory of living cells; the nucleases cleave the phosphate bonds interlinking the nucleotide units. Some enzymes (*phosphodiesterases*) can destroy both DNA and RNA, whereas others (*ribonucleases* for RNA and *deoxyribonucleases* for DNA) are specific for just one type.

Snake venoms, especially those from vipers, possess proteinases which break down a variety of proteins and peptides (Figure 10.4). Some may be non-specific whereas others act on specific substrates like a number of *metal-*

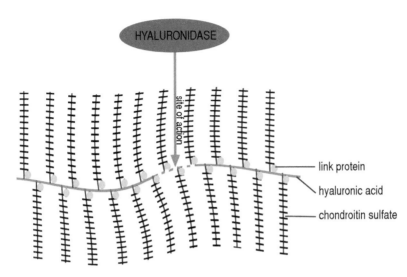

Figure 10.2 The schematic action of hyaluronidase. Proteoglycans are negatively charged polymers which form the ground substance of connective tissues. They determine the viscoelastic properties of joints and other structures. Hyaluronidase destroys hyaluronic acid and hence the general organization of proteoglycans.

loproteases whose activity requires the presence of a metal atom, usually zinc. These enzymes can exert various functions: *collagenases* degrade *collagen*, an important fibrous material, and *elastase* is a digestive enzyme. They also include other enzymes attacking proteins involved in blood coagulation as well as many whose biological roles are as yet unclear.

One enzyme of the many other snake venom enzymes is *L-amino acid oxidase* which degrades amino acids and is present in nearly all snake venoms. It contains a particular sub-fraction, called *flavine adenine dinucleotide* (FAD), which is responsible for the yellowish colour of venoms. FAD is a sort of helper group, which assists the enzymic reaction. The enzyme recognises and transforms almost any natural and free amino acid.

From the above examples, a possible digestive role for venom enzymes follows from their known catalytic reactions. However, this direct relationship might be too simplistic; consider two examples.

1   Certain enzymes which degrade various forms of nucleic acids also decompose a most important compound called *adenosine diphosphate* (ADP) which is known to participate in platelet aggregation. Such enzymes in the venom of vipers and pit vipers inhibit platelet

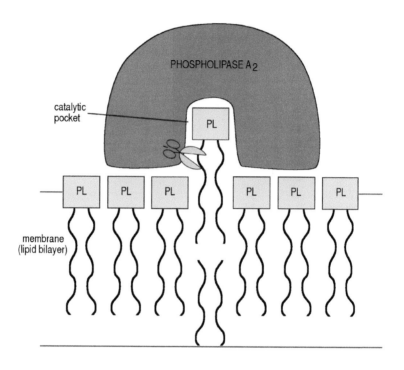

Figure 10.3 The action of a phospholipase A$_2$. Top: a phospholipid (PL) in the membrane is destroyed by a phospholipase A$_2$. Bottom: the chemical reaction that is catalysed by the enzyme. R1 and R2 are long lipidic chains. The real folding of a phospholipase A$_2$ is shown in Figure 11.6.

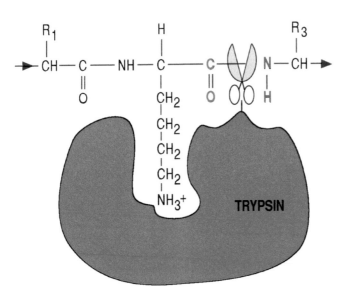

Figure 10.4 Schematic representation of the cleavage of the polypeptide chain of a protein by a specific enzyme. Here, the enzyme is trypsin which selectively cleaves the protein after its positively charged residues lysine and arginine for which trypsin possesses a specific recognition pocket.

aggregation, prevent occlusion and hence repair of damaged blood vessels.

2   The catalytic reaction of L-amino acid oxidase consists of transforming free L-amino acids. When doing so, the enzyme releases not only ammoniac but also generates hydrogen peroxide as a by-product, an extremely reactive oxidising agent which indirectly triggers *apoptosis*, naturally programmed cell death, an important natural process in biological development in which cells no longer needed spontaneously 'commit suicide'. Apoptosis can be identified by a number of behavioural features such as condensation of the cell nucleus and membrane fragmentation. We can see how, through its action on amino acids, L-amino acid oxidase could indirectly favour degradation of the prey's cells.

Another intriguing aspect of snake venoms is that they contain unexpected components. Thus, vertebrates produce small amounts of various factors to regulate the survival of neurones; they are grouped under the general term *neurotrophins*. Some of them, the so-called *Nerve Growth Factors* (NGFs), are non-enzyme proteins composed of two separate amino acid chains each about 120 amino acids long and containing three disulphide bonds (chemical bonds in which two sulphur atoms, either on the same or on different amino acid chains, join together and so stabilise the three-dimensional structure of a protein – see the Appendix for a summary of protein structure). Surprisingly, venoms from elapids and vipers contain substantial amounts of NGF: 1 g of cobra venom includes 1.5 mg of NGF. Nobody knows why; perhaps NGF has some other function unrelated to its recognised primary role to favour nerve growth.

Also striking is the presence of many small compounds in snake venoms. These include lipids, sugars, flavins, a variety of *nucleosides* and *nucleotides* (components of nucleic acids), and amino acids. A number of pharmacologically active compounds like *histamine, serotonin, bufotenin* and *N-methyltryptophan*, found in tissues of many other living organisms, are also present in viper and elapid venoms. *Catecholamines* (neurotransmitter substances) may also occur in some crotalid venoms while acetylcholine (a nerve transmitter substance) is rather abundant in mamba venoms.

Snake venoms also contain a range of positively charged metal ions including sodium, potassium, calcium, magnesium, zinc, sometimes copper and manganese and probably iron. Their presence is explained, at least in part, by the abundance in venoms of enzymes whose catalytic action requires them. However, some of these ions may have other functions. While present knowledge supports the view that 'non-toxic' venom components may have a digestive function, much remains to be done not only to confirm this plausible scenario but also to understand how it is achieved. It would also be of interest to investigate whether the 'non-toxic' components could favour the specific tissue distribution of the 'toxic' components.

11

What are snake toxins?

It is now time to embark on a molecular journey among the venom constituents and especially those that are harmful, the so-called toxins. The word 'toxin' originates from Greek in which *Toxikon* was a 'poison applied on an arrow' as originally defined by people from Scythia, an ancient region of south-eastern Europe on the lower course of the Don and Dnieper. Toxikon is itself derived from *Toxikos*, 'arrows', which in turn comes from *toxon*, 'bow'.

In practice, it is not an easy matter to define a toxin precisely. It is tempting to think of it just as a compound which kills a mouse but this would be very simplistic and incorrect because, as we have already observed, the capacity of a compound to kill depends on a number of factors: the target species, the dose injected, the site of administration and so on. What is toxic under one set of conditions may be harmless under another.

The formal definition of a toxin is a compound which, at low doses, harms or destroys a living organism. Nowadays, however, more and more novel proteins are extracted from venoms and their biological activities are investigated only under *in vitro* conditions. Their capacity, if any, to kill mice or any other types of animals is often not actually tested. Furthermore, with genetic analysis well under way, a number of venom proteins have been identified from their chemical structure as deduced from their gene sequences. It is not clear whether all these venom components should be regarded as toxins.

The synapse: a preferred target for toxins

In 1857 the French physiologist Claude Bernard, wondering what toxins from animals or plants do, reasoned that they should kill by interfering with the proper functioning of central activities essential for life, like the nervous or muscular systems.

To illustrate the line of physiological argument, look briefly at two studies by Bernard involving not snake venom but another poison, curare. This was first brought to Europe from Guyana in 1595 by Sir Walter Raleigh.

It is a complex mixture of strychnine and various other plant extracts. Animals, including large mammals, poisoned with curare are rapidly paralysed and hence easy to catch, so the hunting natives of the Amazonian forests coated their arrows with it.

Bernard found that, to be effective, curare had to be injected into the bloodstream, thereby abolishing the reflex movements of the skeletal muscles and causing paralysis. He poisoned frogs with curare and observed that an electrical excitation of the animals' nerves was not followed by contraction of the corresponding muscles, although direct excitation did cause them to contract. He concluded that, because muscular contraction could occur, although their stimulation via their associated nerves would no longer work, the nervous and muscular systems are distinct from one another. This conclusion implied that there must be some form of communication between a muscle and its stimulating (afferent) nerve; we now call this contact the *neuromuscular junction* or *synapse* (Figures 11.1 and 11.2). We also now know that the curare acts at the synapse to prevent information coming down the nerve from reaching the muscle.

To further localise the action of curare, Bernard carried out another experiment: he selectively blocked the blood vessels supplying the back legs of a frog and introduced curare under the skin in the upper part of the animal's body. That region became paralysed and no longer responded to electric stimulation of the foreleg nerve but, because blood circulation to the back legs was cut off, they were not poisoned by the curare and continued to

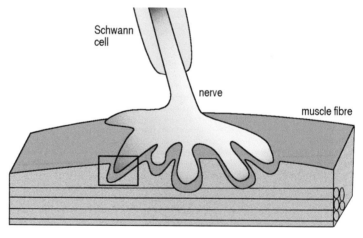

Figure 11.1 The neuromuscular junction. The nerve fibre, wrapped in a Schwann cell, terminates on the muscle. The nerve impulse propagates along the nerve up to the nerve terminal.

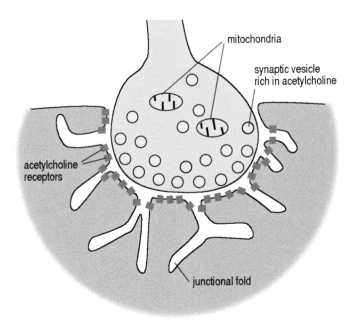

Figure 11.2 A zoom on the small box in Figure 11.1 shows how the nerve terminal enters cavities of the muscle fibre where the postsynaptic membrane is deeply folded. Synaptic vesicles that populate the nerve terminal (about one million in a frog nerve terminal) are filled with acetylcholine (about 10,000 molecules per vesicle). When the nerve impulse reaches the nerve terminal, a special signal (calcium entry) triggers the fusion of vesicles with the presynaptic membrane and, hence, packets of acetylcholine are released in the space between nerve and muscle. The released molecules of acetylcholine diffuse in this synaptic space and reach acetylcholine receptors anchored at the crests of the junctional folds.

respond to electric stimulation of the sciatic nerve. Most surprisingly, however, when Bernard stimulated the nerve in the upper and paralysed region of the body, he observed movement in the hind legs. This experiment not only demonstrated that the motor nerves (those telling muscles what to do) can be distinguished from the sensory nerves (the ones carrying incoming messages of sensation and perception), but also that curare selectively destroys the action of the motor nerves. It is now clearly understood that curare causes respiratory failure and hence death by blocking the action of the motor nerves on the respiratory muscles so that breathing stops and the animal suffocates.

You decide – your brain decides – to move your toe: it takes just one-fiftieth of a second for a nerve impulse to travel from your head to your toe. The rapidity of the information transmission has long been known; 150 years ago nerve impulses had been measured as travelling at speeds of up to about 320 kph (200 mph). The impulse is transmitted so efficiently to muscles by an 'electric wave' propagating rapidly along a nerve and relayed from a nerve ending to its associated muscle by a chemical mechanism which, in turn, propagates another 'electric wave' which travels through the muscle and results in contraction. Transmission of information is first electrical, then chemical and finally electrical again (Figure 11.3).

Sorting out the details of this transmission has been possible over the past three decades, thanks to a wide diversity of toxins from various animals, plants and micro-organisms. Snake toxins have played a major role by helping to clarify how the chemical information emitted by the nerve is received by the muscle and eventually transformed into an electric wave. Let us look at the process from the beginning.

The idea of chemical transmission of nervous stimuli was formulated in 1877 by the German scientist Emil Heinrich Du Bois Reymond. As a natural extension of Bernard's pioneering studies, John Newport Langley provided the first experimental demonstration of this idea using various poisons, such as curare and nicotine. In the absence of any electric stimulation, Langley found that, when applied at exactly the right spot, the place where the nerve enters the muscle, nicotine caused first muscle contraction and then a block. Nicotine therefore behaved first as an *agonist* (an 'activator') and, as the muscle was overstimulated, it *desensitised* (become insensitive to the action of agonists). When curare was applied to the same spot, a muscle responded neither to nerve stimulation nor to the application of nicotine but did react when stimulated directly. Curare behaved as an antagonist of nicotine action. The effects of nicotine and curare persisted even after the nerve was removed. These and many other observations led Langley to correctly conclude that 'the nervous impulse should not pass from nerve to muscle by an electric discharge but by the secretion of a special substance at the end of the nerve'; that special substance is the chemical *acetylcholine*. It is a relatively simple compound released by nerve endings which diffuses across the narrow gap between nerve and muscle and hits the muscle at, as Langley put it, a 'receptive substance', later called the *acetylcholine receptor* (Figure 11.3). This receptor is also the target of nicotine and curare.

In the early 1950s it was uncertain whether the active elements of snake venom were simple or complex. Most, though not all, scientists thought that the main lethal effect of cobra venom was a curare-like paralysis of the respiratory muscles. This was supported by convincing studies performed in 1961 by several Taiwanese scientists and described in 1963. Chuang-Chiung Chang

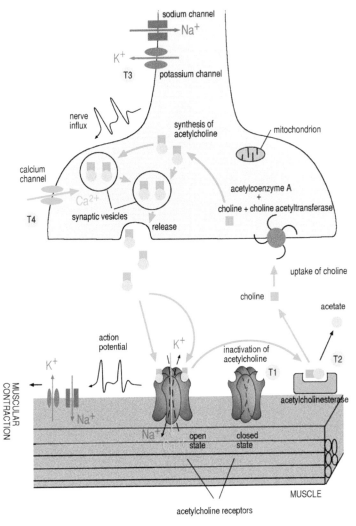

Figure 11.3 A schematic view of the junction of the nerve terminal and the site of action of some snake toxins. The 'life' cycle of acetylcholine is shown, including its synthesis in the nerve terminal, its storage in synaptic vesicles, its release in the synaptic cleft, binding to acetylcholine receptors which thus open and allow ions to go through the membrane, its hydrolysis by acetylcholinesterase into choline and acetate, the uptake of choline in the nerve ending and finally the synthesis of acetylcholine, and the cycle continues. Four sites of action of snake toxins are indicated. T_1 indicates the curaremimetic toxins, T_2 the fasciculins, T_3 the dendrotoxins and T_4 the calciseptine (see text).

and Chen-Yuan Lee, two physiologists in Taiwan, submitted 20 mg of a sample of venom from *Bungarus multicinctus* to electrophoresis (see box, pp. 102–104) in a slab of starch jelly and succeeded in isolating for the first time two different types of snake toxins, the α- *and* β-*neurotoxins*. They looked at these substances in electrophysiological experiments and concluded that α-bungarotoxin blocks neuromuscular transmission by irreversible combination with the acetylcholine receptor in the motor end-plate, whereas the β-bungarotoxins exhibit their action at presynaptic (nerve side) site(s). It was subsequently demonstrated that the curare-like action of all cobra venoms was due to α-bungarotoxin-like compounds.

At first sight, nothing seemed very new, since this snake toxin behaved like curare, which had been known to scientists for a long time. However, the snake toxin had a new characteristic which was essential for the rest of the story. In contrast to curare, whose action vanished rapidly and easily, the block to the acetylcholine receptor lasted so long as to appear irreversible.

Electrophoresis

Modern biochemical methods have enormously simplified analysis of the chemicals present in living systems.

Many of these molecules carry an electric charge and can be made to move in an electric field. The distance any particular molecule travels depends on the strength of the field, the number of charges on the molecule and the molecule's own size: strong fields and high charges make molecules move fast, while large size causes them to move more slowly.

In practical terms, slabs of a jelly, each a few millimetres thick and stiffened with starch or some other polymer, are cast in a mould which leaves a row of pockets at one end. The slab is immersed in a special electrically-conducting liquid and samples of the solutions containing the substances to be analysed are very carefully run into the pockets. When all is ready, the electricity is switched on and the separation is allowed to proceed, often for an hour or more.

After the current has been switched off, various techniques can be employed to visualise the different compounds still in the slab and to see the distance they have travelled. These include dyes, ultraviolet light and other means depending on the substances under investigation. Once a compound of interest has been located, the section of jelly carrying it can be excised from the slab and the compound washed out for further study.

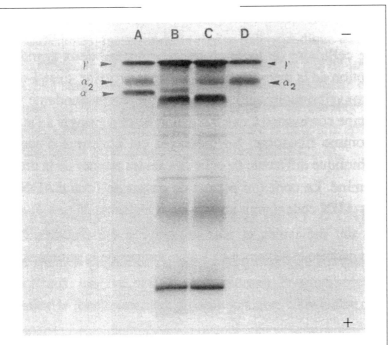

Figure 11.4 Isoelectric focusing in gels of venoms from African spitting cobras and of some of their toxins. The principle of isoelectric focusing is simple. Depending on the pH value, proteins tend to be positively charged, negatively charged or neutral. The pH at which they have no charge is called the isoelectric point (pI). When a protein is submitted to electrophoresis in a pH gradient between an anode and a cathode (the anode is at a lower pH than the cathode), it migrates towards its pI. In this figure, the cathode is at the top of the gel. Lanes A and D show the isoelectric focusing of purified toxins from an African spitting cobra. The cardiotoxin, called γ, is more basic (pI above 10) than the curaremimetic toxin called α2, which is in turn more basic than the curaremimetic toxin called α. These three toxins are present in the whole venom (lane B) which is categorised under the name *Naja nigricollis* by the Institut Pasteur, in Paris, and which was collected some decades ago in eastern Africa, probably between South Egypt and Ethiopia. This venom is quite similar to the venom of a *Naja mossambica pallida* possibly *Naja pallida* nowadays (lane C), collected in Asyut, in South Egypt, and identified by members of the Latoxan company (France) which specialises in the commercialisation of venoms and toxins. Although similar, the two venoms are not identical. The toxin α is absent from the venom collected in Egypt.

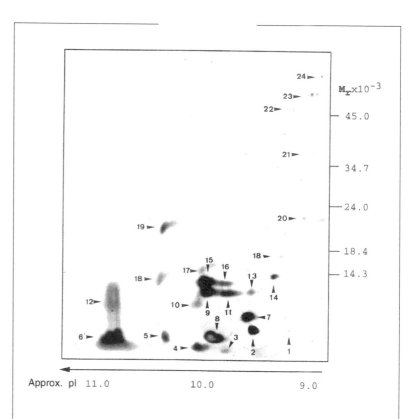

Figure 11.5 Two-dimensional polyacrylamide gel electrophoresis of a snake
venom. Venom components can be separated by methods
which exploit both their charge and their apparent size. The
proteins are first separated by isoelectric focusing in gels, as in
Figure 11.4. Then, the products that have been separated
according to their charge are submitted to an electrophoresis
under denaturing conditions in a perpendicular direction. Typ-
ically, the second dimension consists of a sodium dodecyl sul-
phate (an agent that denatures proteins) polyacrylamide gel
electrophoresis. This procedure leads to a remarkable separation of a complex mixture of proteins such as snake venoms.
This is illustrated here with venom from the sea snake *Laticauda colubrina*. At least 24 spots can be detected. Each of
them was extracted from the gel and submitted to protein
sequencing. The identified succession of the first 10–20 amino
acids of the extracted protein is then compared with the known
protein sequences present in a data bank, and thus most of the
functions of the proteins can be predicted on this simple struc-

tural basis. For example, proteins in spots 9, 11, 15 and 16 are presumably phospholipases A_2, those in spots 2, 4, 5 and 8 are short
curaremimetic toxins, and spot 7 corresponds to a long-chain toxin.
The protein in spot 6 is unknown. The map shown here is sort of a fin-
gerprint of a snake venom. However, considering the great variations
that may occur, not only in a venom throughout the snake's life, but
also from one specimen of a species to another, such fingerprints
should not be used as an identification label of a snake species. Such
a method will certainly help elucidating the 'toxinome', which defines
all the elements that compose the world of toxins and toxin-related
compounds, from animals, plants and microbes.

Indeed, the binding of the toxin to the receptor was so stable that, using a
snake toxin labelled with a radioactive isotope, in the early 1970s Jean-Pierre
Changeux and his co-workers at the Institut Pasteur in Paris identified and
extracted the toxin-receptor complex from its complex muscular context: the
receptor was located using the toxin whose presence was revealed by a specific radioactive label attached to it. The 'receptive substance' postulated by
Langley had finally been discovered: it lies at the junction between nerves
and muscles, is recognised by acetylcholine and by nicotine, and so is called
the *nicotinic acetylcholine receptor* (Figure 11.3). It is a large protein composed of five subunits which crosses the membrane of muscle cells. If you
wish, the receptor looks like a large tooth, a molar say, which crosses the
gum, and has a kind of 'hole' in its centre (see Chapter 12).

Discovery of the acetylcholine receptor helped in the understanding of
how the chemical signal from the nerve is transformed into a contraction
along the muscle. Just under the nerve ending, the muscle possesses a small
region of about one-twentieth of a millimetre in diameter called the *endplate*
which carries about 10 million nicotinic acetylcholine receptors. In the
absence of acetylcholine, the 'hole' in each receptor is closed and thus the
receptor is resting. Upon binding to acetylcholine, receptors open for a thou-
sandth of a second or so, allowing positively charged sodium, potassium and
calcium ions to pass through, with remarkable consequences for the electrical properties of the muscle membrane.

At rest, the distribution of ions is uneven across the membranes of
nerves and muscles. The concentration of sodium, calcium and chloride ions
is much higher outside the cell, with more potassium ions inside the cell. In
the resting state, therefore, there is a difference in the electrical charges
between the inside and outside of the cell, with a potential difference of
about $-80\,mV$ across the membrane. The membrane that separates the two
different charges is said to be *polarised*. The 'opening' of the receptor hole

allows a flow of sodium and calcium ions from outside to inside the muscle membrane and an efflux of potassium ions, all the ions going through the same channel. Such a flow causes a local short circuit across the membrane and a reduction of the initial potential difference almost to zero; the membrane thus becomes *depolarised*. In molecular terms, a nerve impulse triggers the release of 50 to 3,000 packets (*quanta*), each containing about 10,000 molecules of acetylcholine. The chemical transmitter diffuses rapidly (a few microseconds) the short distance that separates the nerve from the muscle (about 0.00005 mm) and causes between 100,000 and 600,000 receptor molecules to open transiently, producing a short circuit. This almost depolarises the endplate locally, abolishing the original voltage difference that existed across the muscle membrane.

This local phenomenon does not, of course, explain how the whole muscle fibre contracts: opening the receptor just triggers a process which then spreads along the fibre. Depolarisation of the endplate also depolarises a substantial surrounding area of the muscle fibre because this region is also full of channels allowing ions to cross the membrane. These channels are not sensitive to acetylcholine but are triggered by a voltage change like the one at the endplate; they are sensitive to the voltage change generated by opening the nicotinic acetylcholine receptors at the endplate. Two major voltage-gated channels are spread along the muscle fibre, the sodium and potassium channels which allow their respective ions through when they open. Thus, voltage-gated channels relay the chemical depolarisation initiated by the nicotinic acetylcholine receptor and this propagates an electrical wave along the muscle fibre. A similar voltage difference-based process explains the propagation of an impulse along a nerve.

The discovery of the nicotinic acetylcholine receptor did not close the neuromuscular transmission story but heralded a new era during which many other receptors, ion channels, enzymes, and such involved in transmission were discovered by the same basic approach. Most clearly, toxins were exceptional tools for the molecular dissection of the essential elements of life.

Blood – important target of snake toxins

Viper and other snake venoms often affect blood by interfering with *haemostasis* and *fibrinolysis*, two complex words that hide simple concepts. Blood is not an immobile liquid: its flow is vital for the transport of various essential materials, including oxygen, around the body. Blood flow and the vascular system's self-repair mechanisms, like the formation or destruction of clots, clearly need to be adequately regulated. Haemostasis comprises all of the reactions activated on vascular injury to repair the damage and so prevent blood loss. This process basically involves three steps:

1   a spasm or constriction reflex of the vessel wall with stimulation of a number of chemical factors;
2   aggregation of platelets (small cells found in blood which function specifically as blood clotting centres) with the formation of a nucleus around which the clot develops. The platelets release substances that help prolong the spasm and attract more platelets;
3   a solid clot associated with a fibrin network encloses the original mass of platelets and blood coagulates. Coagulation, the transformation of liquid blood into a gel, is only one aspect of haemostasis which blocks the punctured blood vessel and thus stops the bleeding.

The clot has to be destroyed if blood is to flow freely again. This is done by *fibrinolysis* (meaning 'breakdown of fibrin'), in which various proteins from the tissues, blood cells and plasma interact to degrade the clot. More than 70 different proteins have been identified in human plasma and at least 25 of them are involved in the haemostatic and fibrinolytic systems. Alteration of any of these proteins may cause serious damage, including copious bleeding, haemorrhage or clotting which occludes a blood vessel or a heart cavity.

The architecture of snake toxins
We are now entering the real molecular world in which we shall need to employ a number of basic notions regarding protein chemistry. These are really quite simple, but the methods used to study the proteins are, however, complex; the Appendix has a brief summary of protein structure and chemistry. Since the discovery of α-bungarotoxin, many other toxins have been found in snake venoms. What is the structural basis for the diversity of toxic functions in snake venoms? Is each toxin architecturally unique or do toxins share major features? During the past two decades, detailed snake toxin architectures have been worked out to reveal a most striking phenomenon: there are more biological functions than there are toxin structures. Thus, toxins acting on different targets may have similar architectures and patterns of folding: the folds adopted by three enzymes and a number of non-enzymic proteins, will now illustrate the principles.

Enzymic proteins
The *phospholipases* A_2 are well known enzymes catalysing the splitting of the phospholipids into lysophospholipids and fatty acids. The phospholipase A_2 fold is shown in Figure 11.6. It is a protein with a single chain of about

120 amino acids and seven disulphide bridges, a good deal of helical structure, two strands of β-pleated sheet and several turns (see Appendix). Snake toxins with such a fold may perform a variety of toxic functions which sometimes require the enzymic activity and sometimes not.

- The presynaptic toxins (β-neurotoxins) adopt the phospholipase A₂ fold. They were discovered in 1963 by Chang and Lee when they separated the components of krait venom. Since then, several forms of this toxin (β1-bungarotoxin, β2-bungarotoxin, etc.) have been found in the venom: each is highly toxic since an intravenous injection of 400 nanograms (one 2.5 millionth of a gram) of β1-bungarotoxin can kill a 20 g mouse. Various elapids and vipers also produce toxins like this: the most potent β-neurotoxin so far known is *textilotoxin* from the venom of a snake called *Pseudonaja textilis*; it takes just 20 nanograms (one 50 millionth of a gram) to kill a mouse.

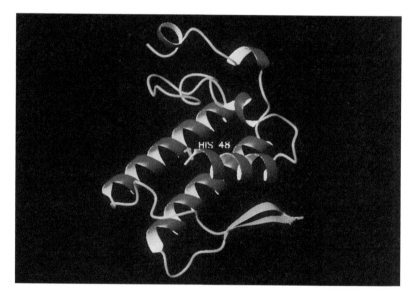

Figure 11.6 Architecture of *Phospholipase A₂* fold that exerts multiple toxic functions in snake venoms. For the sake of clarity, only the backbone structure is shown, the side chains of the amino acids having been omitted. The regions of the protein that adopt a β-sheet structure are coloured orange. Red and yellow indicate regions that adopt an α-helix structure. The particular structure shown corresponds to the X-ray structure of the potent notexin from the Australian Tiger snake, *Notechis scutatus scutatus*. It was solved by B. Westerlund *et al.*, in 1992. Histidine 48 is essential for the enzymatic activity of notexin.

The presynaptic toxins exert a potent action on the nerve ending, causing a failure in the release of acetylcholine which generally occurs in three phases. Transmission is initially suppressed, then enhanced and finally declines to complete failure. As a consequence, the nerve emits no signal and no muscle contraction can occur: β-neurotoxins cause flaccid paralysis of skeletal muscles and death results from respiratory failure due to paralysis of the diaphragm.

The target of β-bungarotoxins seems to be voltage-gated potassium channels involved in the propagation of the nerve impulse. Several groups of scientists are searching for the receptors of the other presynaptic toxins which may not be potassium channels.

The simplest presynaptic toxins involve just a single phospholipase A_2 motif, as in the case of the potent *notexin* from tiger snake venom. Other β-neurotoxins act in partnership with one or more other proteins (Figure 11.7). For example, the basic phospholipase A_2 motif in crotoxin from the South American *Crotalus durissus terrificus* is associated with another phospholipase A_2 motif whose protein chain, however, is acidic and split into various parts. β-Bungarotoxins also possess two protein motifs linked by a disulphide bond, the additional protein adopting a 'pear' fold (see Figure 11.12). The potent *taipoxin* from the Australian elapid snake Taipan (*Oxyuranus scutellatus scutellatus*) involves three phospholipase A_2 motifs, one of which is much more toxic than the other two. The even more potent *textilotoxin* from another snake includes five phospholipase A_2 motifs, two identical and three different. The basic phospholipase A_2 motif can therefore be involved in a range of combinations: apparently the greater the complexity the greater the toxicity but it is, for the moment, unclear how additional motifs increase the toxic activity of the molecule.

• The phospholipase A_2 motif has many other functions. Some *myotoxins* (muscle poisons) affect muscles remote from the site of poison injection, causing muscle degeneration and excretion of *myoglobin* (the red-coloured oxygen-transferring protein of muscle) in urine, a sign of renal failure. The venom of the sea snake *Enhydrina schistosa* contains such a myotoxin. Others act on the muscle locally, such as some toxins from *Bothrops atrox* venom, right at the injection site.

The molecular mechanisms underlying the myotoxic actions of snake venom phospholipases A_2 remain unknown. More unexpectedly, *notexin* from the Australian tiger snake is not only a presynaptic toxin but also a myotoxin. This venom phospholipase A_2 thus displays at least two biological functions, but the value to the tiger snake of such a polyvalent toxin is unknown.

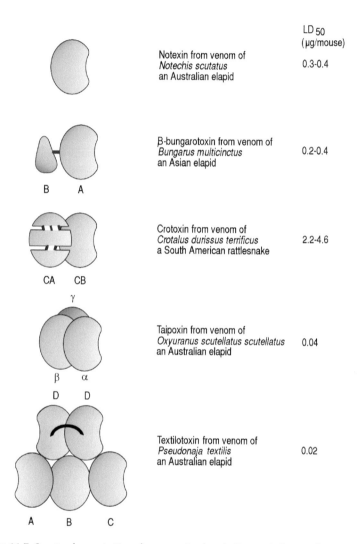

	LD 50 (µg/mouse)
Notexin from venom of *Notechis scutatus* an Australian elapid	0.3-0.4
β-bungarotoxin from venom of *Bungarus multicinctus* an Asian elapid	0.2-0.4
Crotoxin from venom of *Crotalus durissus terrificus* a South American rattlesnake	2.2-4.6
Taipoxin from venom of *Oxyuranus scutellatus scutellatus* an Australian elapid	0.04
Textilotoxin from venom of *Pseudonaja textilis* an Australian elapid	0.02

Figure 11.7 Structural organisation of presynaptic phospholipases A_2 from snake venoms. At least one phospholipase A_2 fold, indicated by a bean-like structure, is found in all toxins. In the β-bungarotoxin, the phospholipase A_2 fold is linked covalently to a small protein adopting a pear fold (see text). In crotoxin, it is non-covalently linked to pieces of a phospholipase A_2. Three and five phospholipases A_2 domains are found in taipoxin and textilotoxin, respectively. Note that the quaternary (spatial) organisation of the different partners has been defined arbitrarily. LD_{50} is the lethal dose of toxin which, injected intraveinously (iv), kills 50 per cent of a group of 20g mice. The variations in LD_{50} illustrate differences in measurements from one laboratory to another.

- Some phospholipases A_2 may destroy various cells and are called 'cytotoxins'. *Nigexine* is such a cytotoxin from the venom of an African spitting snake. It causes fatal haemolysis by destroying red blood cells and this lethal haemolytic action requires the enzymic activity. Nigexine, however, can do more: it is capable of destroying tumour cells but in this function it does not require the enzymic activity. By abolishing just the enzyme function, the haemolytic activity (and hence the lethal property of nigexine) is suppressed while retaining the capacity to destroy tumour cells. That might be of interest for the treatment of some tumours. Other phospholipases A_2 have anticoagulant properties, can initiate or prevent platelet aggregation, induce convulsions, hypotension or oedema.

The serine proteases are enzymes which, like trypsin and chymotrypsin, possess a catalytic machinery. It involves three crucial amino acids (aspartic acid, histidine and serine) which act together to break the chains of proteins. The serine protease fold characterises a single protein chain of about 230 amino acids with six disulphide bridges (Figure 11.8).

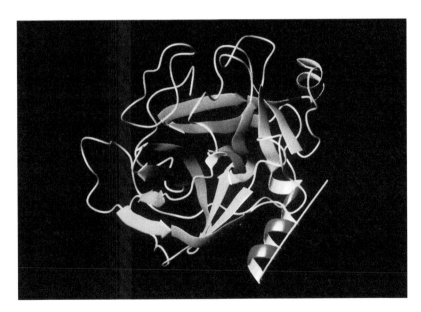

Figure 11.8 Architecture of the backbone of the *Serine protease* fold. The particular structure is that of the plasminogen activator, called TSV-PA, isolated from *Trimeresurus stejnegeri* venom. Its X-ray structure was solved by Parry *et al.* in 1998. Serine 195, which is essential for the catalytic action of the enzyme, is shown in cyan.

Snake venom serine proteases usually interact with proteins involved in haemostasis and fibrinolysis. Thus, serine proteases from Russell's viper venom are coagulants which activate one of the natural factors assisting coagulation, whereas proteases from elapid venoms activate another such factor. Coagulating serine proteases, found exclusively in Australian snake venoms, are prothrombin activators while those from various viper venoms are thrombin-like enzymes, converting fibrinogen into fibrin clots. Other serine proteases can act as anticoagulants, again through different modes of action.

Like the phospholipase A_2 fold, the serine protease fold has been intensively exploited by snakes to generate a multiplicity of biological functions, mostly directed this time to the transformation of blood.

The zinc metalloproteases are other enzymes with multiple biological functions. The name 'zinc metalloprotease' means an enzyme which needs at least one zinc atom to break down proteins. The zinc metalloprotease fold in snake venoms possesses about 200 amino acids with a zinc-binding site around the end of the chain, essential for the fold to act as an enzyme and to be able to cleave other proteins (see Figure 11.9). Some metalloprotease domains from snake venom can be followed successively by a small spacer region of about 20 amino acids and a *disintegrin* domain containing 40–80 amino acids which is known to bind to integrin, a protein present on platelet membranes which is normally recognised by fibrinogen. In other cases, this succession of three domains may be followed by a domain of uncertain function rich in sulphur-containing amino acids as, for example, in the toxin *Jararhagin* from *Bothrops* venom. Some even longer toxins may possess an additional domain which binds to certain plant proteins. This means that the size of a snake venom protein containing a metalloprotease fold may vary from about 200 amino acids to almost 1,000. Such a complex structure may exert a variety of biological activities.

Most snake venom metalloproteases produce severe haemorrhage or bleeding. They also often inhibit formation of a platelet plug (which reduces the effectiveness of staunching blood loss from a ruptured vessel) and degrade basement membranes and matrix components outside cells. Others inactivate the protease inhibitors present in plasma called *serpins*, abolishing the protective effect of serpins against proteolytic activity in blood. *Mocarhagin*, from the venom of an African cobra, is a little peculiar in that it acts as an anti-platelet aggregation factor, destroying the receptor at the surface of platelets that are crucial for aggregation.

In conclusion, at least three protein folds with different enzymic activities are largely exploited by snakes to exert potent toxic activities. Curiously, the enzymic property is not always necessary for the toxic action to be

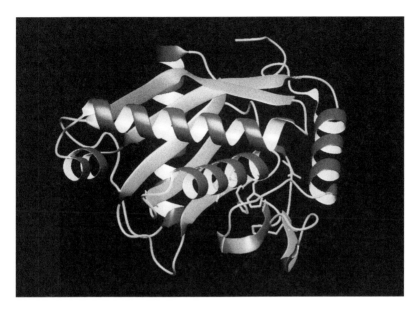

Figure 11.9 Architecture of the backbone of the *Metalloprotease fold*. This particular structure is that of atrolysin C, a potent haemorrhagic toxin from the venom of the western diamondback rattlesnake *Crotalus atrox*. Its X-ray structure was solved by Zhang *et al.* in 1994. Histidine residues 146, 146 and 152 coordinate the catalytic zinc and are coloured cyan.

expressed, suggesting that it is primarily the framework of the enzyme that is essential for developing new toxic functions. This conclusion tallies well with the observation that protein folds with no enzymic activity can also exert quite diverse toxic activities.

Folds with no enzymic activity
The three-finger fold is present only in elapid venoms and the reason for the fold's name is clear from Figure 11.10. It is a simple chain of about 60–74 amino acids which folds up into three adjacent loops or 'fingers', hanging down from a compact region of the protein where four disulphide bonds are located. Its major secondary structure is a large β-pleated sheet with several turns but no helical structure (see Appendix).

Three-fingered toxins can be divided into at least four subgroups:

1 short toxins with about 60 residues and four disulphide bridges;
2 long toxins with about 70 residues and five disulphide bridges. The fifth bond of long toxins is usually located at the tip of the central loop, forming a small extra turn;

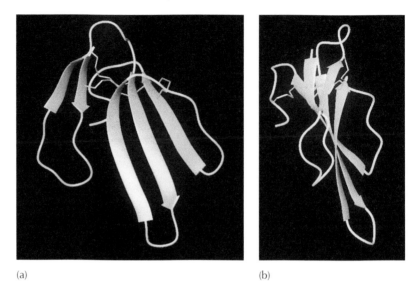

(a)   (b)

Figure 11.10 The structure of a snake toxin that adopts a *three-finger fold*. For the sake
of clarity, the side chains of the amino acids have been omitted. The
backbone regions of the toxin that adopt a β-sheet structure are coloured
orange. (a) the structure is shown from the concave face. (b) the same
structure turned 90° to the right. This reveals the flatness of the three-
finger fold, with its two major faces, one of which is in contact with the
end of the backbone that forms a terminal loop (on the left of the structure shown on the right). The X-ray structure shown is that of erabutoxin
a, from the sea snake, *Laticauda semifasciata* which was elucidated by
Saludjian *et al.* in 1992. Also shown are the four disulphide bonds that
are located in the upper part of the structure.

3   a recently discovered subgroup in venoms from Asian kraits, Asian
    cobras and South American coral snake includes long toxins whose extra
    bridges are located on the first finger.
4   a few toxins have a 'hybrid' structure with the number of residues typical
    of long toxins but with only four disulphides.

The three-fingered fold can thus encompass many variations, including a
short first loop and a long tail in long toxins and, an extra disulphide bond at
the tip of the second loop or on the first loop. With so many possible structural variants, this fold can have many functions (Figure 11.11) which we
shall now look at in detail.

*   *Curaremimetic toxins or α-neurotoxins* are potent toxins (1–4 millionths of
    a gram may kill a 20 g mouse) produced in high concentrations in some

venoms. Thus, they constitute up to 70 per cent of all proteins from the venom of sea snakes of the genus *Laticauda*. Like α-bungarotoxin, all curaremimetic toxins block the nicotinic acetylcholine receptor, preventing transmission of nerve impulses to the skeletal muscles (including the diaphragm) which remain paralysed; death results from respiratory failure. The ultimate effects of their action are similar to those of β-neurotoxins but the primary modes of attack are radically different. Blockage of nerve impulse transmission to the muscle is achieved by β-neurotoxins which prevents the nerve from releasing acetylcholine, and by α-neurotoxin which stops acetylcholine from reaching its target. A venom with both types of toxin will immediately generate a strong, long-lasting and dangerous muscle paralysis, exactly what happens with many elapid venoms.

More than 100 curare-like toxins have been extracted from venoms of various terrestrial and marine elapids. All show comparable curare-like actions; about 50 are short toxins with 60 to 62 amino acids, while more than 40 others are long and have 66 to 74 amino acids together with five or four disulphide bonds.

- Some three-finger toxins bind to other nicotinic acetylcholine receptors present in neurones of vertebrates. Found uniquely in krait venoms, these are *K-neurotoxins* which bind to receptors possessing two types of subunits present in the brain and in various ganglia. Whilst curaremimetic toxins consist of a single three-finger motif, K-neurotoxins are composed of two identical motifs of a long-chain-type three-finger fold which are bound to each other.
- *Muscarinic* toxins are short chain three-finger toxins, present exclusively in mamba venoms. Their name describes their action on muscarinic acetylcholine receptors present in brain and various tissues of vertebrates. Although muscarinic and nicotinic acetylcholine receptors are both activated by acetylcholine, they are structurally and functionally unrelated. The nicotinic species are channels composed of five transmembrane subunits, whereas the muscarinic receptors are non-channel membrane proteins resembling other types of receptors which all possess seven helices across cell membranes. Muscarinic receptors have a binding site for acetylcholine accessible from the extracellular side of the membrane; when acetylcholine binds, the receptor becomes capable of binding a particular protein located inside the cell and this, in turn, triggers a number of intracellular processes including opening of pores in other proteins. These receptors can regulate heart rate and contraction and relaxation of blood vessels, and may influence cognitive processes. Several muscarinic toxins have been isolated from mamba venoms; some

short chain curaremimetic toxin
toxin α from *Naja nigricollis*
NMR Zinn–Justin et al.,1992

long chain curaremimetic toxin
α-cobratoxin from *N. kaouthia*
RX Betzel et al., 1991

neuronal toxin
k-bungarotoxin *B. multicinctus*
RX Dewan et al., 1994

acetylcholinesterase blocker
fasciculin 1 from *D. angusticeps*
RX Le Du et al., 1992

muscarinic receptor ligand
MTX2 from *D. angusticeps*
NMR Segalas et al., 1995

(a)

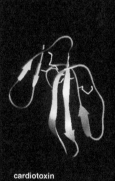

cardiotoxin
toxin γ from *Naja nigricollis*
NMR Gilquin et al.,1993

Ca2+ channel blocker
toxin FS2 from *D.p. polylepis*
NMR Albrand et al., 1995

cell adhesion protein
mambin from *D.j. kaimosae*
NMR Sutcliffe et al., 1994

(b)

are simple blockers that prevent binding of acetylcholine while, most unexpectedly, others are *agonists* which behave like acetylcholine and *activate* the receptors.

- *Fasciculins* are short three-finger toxins also found only in mamba venoms. Although they seem to cause death by respiratory paralysis, they are weakly toxic and it may take 50 or 60 times more than a curaremimetic toxin to kill that unfortunate mouse. When injected, mice display long-lasting fasciculations (rapid contractions of various muscles, like local shiverings) that wane after a few hours. This results from an inhibition of acetylcholinesterase (see Figure 11.3), an enzyme which plays a key role in the regulation of acetylcholine in the synaptic cleft. With acetylcholinesterase blocked, acetylcholine cannot be destroyed so its concentration at the neuromuscular junction remains high and augments neuromuscular transmission by increasing the amplitude and duration of end-plate potentials.

- *Cytotoxins* (a general word meaning substances which can poison or destroy cells) are uniquely found in venoms of terrestrial elapids, particularly cobras and Ringhals. They are moderately toxic (about five times less than the curaremimetic toxins), short-chain type three-finger toxins with 60 amino acids. Also called *cardiotoxins* because they can produce profound cardiovascular depression, they cause the breakdown and dissolution of various tissues, including red blood cells and foetal amnion cells, and may induce protein aggregation and depolarisation of various excitable cells. Details of their modes of action, and especially of their targets, remain obscure: membrane phospholipids may be their primary

---

Figure 11.11 Structural variations around the three-finger fold. Eight different toxins which all adopt a three-finger fold are displayed. (a) Top left: a short-chain curaremimetic toxin (60–62 amino acids and 4 disulphide bonds), a blocker of muscular acetylcholine receptors only. Top middle: a long-chain curaremimetic toxin (66–74 amino acids and 4 or 5 disulphide bonds) which blocks both muscular and some neuronal (α7) acetylcholine receptors. Top right: A neuronal neurotoxin that binds to other neuronal acetylcholine receptors. Unlike the curaremimetic toxins which act as monomers, to be active the neuronal toxin must be in the form of a dimer (combination of two monomers). Bottom left: A blocker of acetylcholinesterase, an enzyme which degrades acetylcholine. Bottom right: A muscarinic toxin that acts on muscarinic acetylcholine receptors. (b) Left: the structure of a cardiotoxin from venom of *Naja nigricollis* (Pasteur Institute). Middle: a blocker of calcium channel. Right: Mambin, a cell adhesion protein which inhibits platelet aggregation. Note how the tips of the fingers are differently twisted from one toxin to another.

target but a specific receptor protein on cell membranes cannot be excluded.

- *Calciseptine* is another short three-finger toxin of 60 amino acids which recognises channels selective for calcium ions. This toxin is also found only in mamba venom; its toxicity in mice is unknown.
- *Mambin* is another toxin confined to mamba venoms. It is a peculiar short three-finger toxin which inhibits platelet aggregation by binding to integrin. When fibrinogen binds to integrins on the platelet surface, a network of platelets forms, plugs the damaged vessels and stops blood loss. Mambin acts as an antagonist, preventing fibrinogen binding to its receptor and hence inhibiting formation of the platelet plug: in the presence of mambin, blood loss continues.

Snake venoms contain many other three-finger toxins whose functions have not yet been elucidated. A number have been isolated and their primary structures are known; they are certainly three-finger proteins but it is not yet clear what they do and what their targets are.

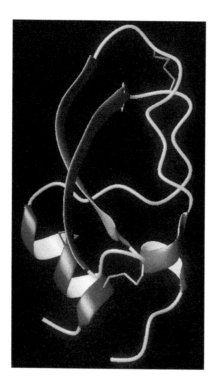

Figure 11.12 *The pear fold.* This name alludes to the fold's pear-shaped architecture. Two toxins from mamba venoms are known to adopt this fold. These are the dendrotoxins which promote release of acetylcholine from nerve endings by blocking voltage-gated potassium channels and calcicludine which acts on calcium channels. The X-ray structure shown is that of α-dendrotoxin, which was solved by Skarzynski in 1992.

The *'pear' fold* is yet another structure found in mamba and other venoms (Figure 11.12). It comprises about 60 amino acids, three disulphide bonds, two strands of β-sheet, two short helices and several turns. Only two functions have so far been ascribed to it. One is displayed by dendrotoxins which bind to voltage-gated potassium channels at nerve endings and cause acetylcholine release. A family of dendrotoxins has been identified, some of which preferentially bind to certain potassium channel subtypes. On intravenous or intraperitoneal injection, dendrotoxins are weakly toxic but are 10,000-fold more potent when injected intracerebrally. They contribute with fasciculin to increase the amount of acetylcholine at the nerve–muscle junction. Dendrotoxin and the three-finger fasciculin are more toxic when acting together than when each acts alone. The two toxins may act in synergy. Another toxin with a 'pear' fold is called *calcicludine* and blocks calcium channels. Many protease inhibitors adopt the same fold.

Other non-enzymic folds have been identified in snake venoms but so far only one function has been associated with them. However, just 20 years ago only two functions were known for the three-finger fold and nobody knew that mamba venoms include pear fold toxins. Looking forward, therefore, one can reasonably predict that, in the next couple of decades, many other activities will be identified for these folds known at present for a single activity. Among them are:

- *waglerins* found in pit viper venom. They are small peptides of about 22 amino acids and one disulphide bond which block acetylcholine receptors at the neuromuscular junction. They compete with the elapid curaremimetic toxins for these receptor sites;
- *disintegrins*, also present in viper venoms. They are single-chain proteins each with 47 to 83 amino acids and four to seven disulphide bonds (Figure 11.13). Like *mambin*, they inhibit platelet aggregation but structurally are quite unrelated, so vipers and mambas have arrived at a similar form of activity by different routes;
- *myotoxins*, containing about 45 amino acid residues and three disulphide bonds, which occur in crotal snakes; they cause muscle damage but quite how is unclear;
- *sarafotoxins*, which are highly potent toxins from atractaspid venoms, with 21 amino acids and two disulphide bonds (Figure 11.14). Binding to the same receptors as endothelins, they are potent vasoconstrictors of the aorta and coronary blood vessels and cause coronary insufficiency resulting from acute vasospasm. Structurally they consist of a small helix, a turn and an extended strand.

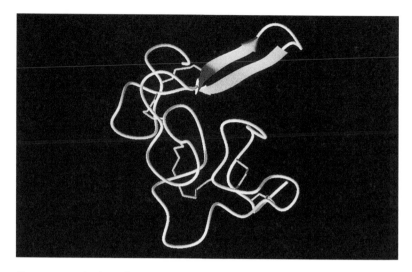

Figure 11.13 The three-dimensional structure of a disintegrin (Kistrin) from the venom of a viper, solved using NMR spectroscopy by Adler *et al.*, in 1991.

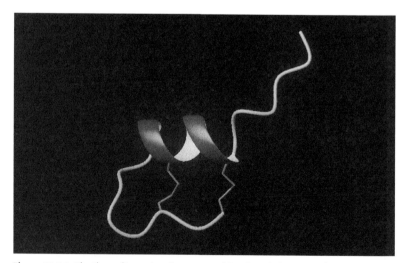

Figure 11.14 The three-dimensional structure of Sarafotoxin b, a toxin from *Atractaspis engaddensis,* solved by Smith *et al.*, in 1995.

Toxin architectures are everywhere, not just in venoms

Toxin folds appear well adapted to multiple biological activities, especially alteration of the prey's functioning physiological systems. However, they are by no means exclusive to snakes but are present in many tissues other than venom glands where they have other biological functions. Thus, the folds with enzymic activity include the phospholipase A_2 fold present in the pancreatic juices of all vertebrates (including mammals and reptiles), where it has a digestive function. The serine proteinase fold is found in a range of digestive enzymes. Finally the zinc metalloproteinase fold is also present in mammalian reproductive proteins from sperm and at the surface of mammalian macrophages. Curiously, although they possess a metalloprotease domain, these proteins exert no known proteolytic activity.

The folds with no enzymic activity are also impressively ubiquitous. In particular the three finger fold is there in a small plant *lectin* which has a high affinity for sugar residues and in CD59, a protein at the surface of various cells where it exerts a defensive function by inhibiting *complement*-mediated destruction of cells (the complement system is a group of blood proteins involved in the destruction of foreign cells after they have been coated with antibody as part of an immune response to the invasion). Perhaps more surprisingly, an American group recently discovered a protein containing a three-finger fold in the brain of a mammal which seems to modulate the natural action of acetylcholine receptors. Does this mean that we have toxins in our brain? Again, we should recall the dictum of Paracelsus: 'What is there that is not a poison? . . . Solely the dose determines that a thing is not a poison.' It is clear that if the newly discovered protein does not exert a toxic action, it certainly possesses a toxin structure.

Each toxin fold has structural equivalents in other tissues which suggests that all these proteins share a common ancestor – but what this common ancestor has been is still shrouded in mystery.

12

Toxins in action

How do snake toxin folds work? As we noted earlier, the folds correspond only to protein backbones, a framework which in reality displays amino acid residues with their variety of side chains (see Appendix). To understand toxin action, we must identify the residues of the fold which come into contact with the target molecule: as we will shortly see, no more than a small portion of exposed residues on the toxin are directly implicated in its function.

How does a snake toxin bind to its target?

Consider the simplest situation of a non-enzymic snake toxin acting as an obstruction, sticking onto the receptor and so preventing the attachment of the natural transmitter. This is the case with the sea snake curaremimetic toxin *erabutoxin a*, whose binding to the acetylcholine receptor precludes binding of acetylcholine (Figure 12.1).

The 'toxic' topography of erabutoxin a has been elucidated by an approach called *mutational analysis*, which consists in changing, one at a time, most of the amino acids of the toxin and then investigating the biological consequences of each such change. To do this, a way first had to be found of producing each variant toxin artificially; the snakes themselves generate just a few variants of a toxin – the ones that actually work! This is where modern techniques of genetic engineering came into play. The gene for a toxin, isolated from the snake's DNA, was inserted into the genetic machinery of a suitable bacterium, like the well-known *Escherichia coli*. The bacterium recognised the novel gene as its own and produced the toxin, just like the snake does. The production factory working properly, the original gene then had to be modified in a predetermined manner. This was done using a great variety of commercialised enzymes, after which the modified genes were inserted into the bacterium which produced toxin molecules carrying the imposed changes and which could be tested for toxicity. In this way, the amino acids in the toxin protein essential for its toxic action could be distinguished from those that are not. Moreover, the 20 g mice are now left in peace; nowadays we usually measure toxicity not in whole animals but by

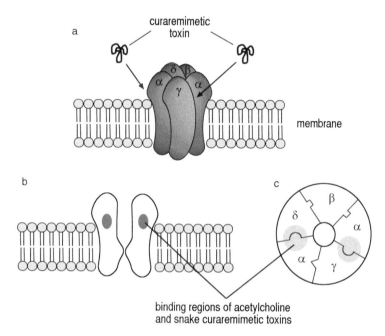

binding regions of acetylcholine
and snake curaremimetic toxins

Figure 12.1 A schematic representation of the muscular-type nicotinic acetylcholine
receptor in its membrane. The five subunits of the receptor, two α, one β,
one γ, and one δ are shown in (a). Longitudinal and transversal sections of
the receptor are in (b) and (c) respectively. The receptor regions where the
natural transmitter, acetylcholine, and the snake curaremimetic toxins
bind are indicated by the red spots. They are located at the two interfaces
between the subunits αδ and αγ.

determining in an artificial system the strength (affinity) of binding of the
mutant toxin to its isolated target, in this case the nicotinic acetylcholine
receptor. More precisely, we search for the residues by which erabutoxin a
binds to the acetylcholine receptor present in the electric organ of an electric
ray, *Torpedo marmorata*. The reason for this choice is two-fold. First, the
acetylcholine receptor abounds in electric organs of rays and can be compared with the receptor present in the neuromuscular junctions of most vertebrates. Second, since sea snakes feed on fish, we expect that the binding
properties of the sea snake, erabutoxin a, will be similar to the receptors of the
prey.

The structure of erabutoxin a is rich in β-pleated sheet that encompasses
its three fingers. The sheet defines a flat plane with two faces differentiated
by the position of a small loop closed by a disulphide and involving the last

residues of the fold (Figure 11.10). This small loop is always in contact with the same face of the plane. The strength of binding of erabutoxin a to the acetylcholine receptor from an electric ray substantially decreased when a particular set of ten of its amino acids (residues in orange and red) were mutated; mutations at the other amino acid positions had very weak (residues in yellow) or no (residues in green) effect on the toxin's potency. The side chains of the ten mutation-sensitive residues are clearly those by which the toxin binds to its target. They are essentially displayed on the face that is not in contact with the small loop, spread on the three fingers and thus forming a large homogeneous surface (Figure 12.2). Not all the functional residues are

Figure 12.2 The 'curaremimetic' site of erabutoxin a. The toxin is represented as in Figure 11.10, its three-finger planar face being viewed from the left and turned 90° to the right. In contrast to Figure 11.10, however, the toxin is shown with all its amino acid side chains, thus masking the toxin backbone. The residues that contribute to binding to nicotinic acetylcholine receptor from the electric organ from the fish *Torpedo marmorata* are coloured red for the most important residues, orange for the moderately important ones and yellow for those that are not directly important but are located at the border of the 'toxic' region. The only residue whose change from the original amino acid (an isoleucine, replaced by arginine) caused a substantial increase in receptor-binding strength is shown in blue. In other words, the toxicity of erabutoxin a increases upon this single amino acid change. The 'toxic' residues on the first loop are glutamine 7, serine 8 and glutamine 10. Those on the central finger are lysine 27, tryptophan 29, aspartic acid 31, arginine 33, isoleucine 36 and glutamic acid 38. Only lysine 47 is located on the third loop. The view on the right shows that the 'toxic' residues are located only on the face (right of the figure), opposite the terminal loop. In a sense, the back of the toxin is 'harmless'.

equally important for binding; those in red are more important than those in orange, but one seems particularly critical: a positively charged arginine (R33) at the tip of the central finger. Though less critical, two positively charged lysines on the second and third fingers are also important. It is probably the binding of this 'toxic' topography to the nicotinic acetylcholine receptor that engenders the harmful effect of erabutoxin a. The complementary surface offered by the receptor has now been identified (see Figure A5).

If we destroy the three-finger fold of erabutoxin a by cleaving its four disulphide bonds (there are specific reagents for this), the toxin loses its capacity to bind to the receptor. The reason is simple. When the fold is destroyed, the functional amino acids are well separated along the extended chain. They are not close together in space and hence do not form the required homogeneous 'toxic' topography. The 'toxic' surface therefore is closely associated with the toxin fold (Figure 12.3).

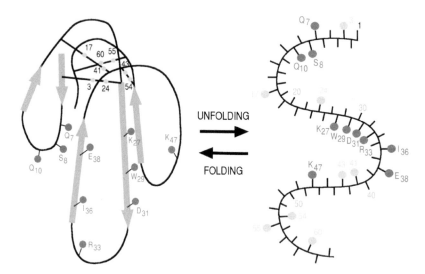

Figure 12.3 Folding and unfolding of a three-finger toxin. Left: the folded structure of erabutoxin a, with the 10 'toxic' residues (red), the four disulphide bonds (green), the five β-sheet strands (long arrows in blue) and the three fingers. Right: upon breakage (reduction) of the disulphide bonds by specific reagents, the whole architecture collapses. Thus, the 'toxic' residues do not all remain in spatial proximity but are spread along the chain. See, for example, Q_7 and E_{38}. The unfolded molecule is no longer a potent toxin.

Not one but many 'toxic' sites

Identification of the 'toxic topography' of the long toxin α-cobratoxin from an Asian cobra showed that not all curaremimetic toxins use exactly the same topography as erabutoxin a to bind strongly to the same acetylcholine receptor (Figure 12.4). The three fingers are implicated in the contact between the short toxin and the receptor but, with the long one, only residues of the

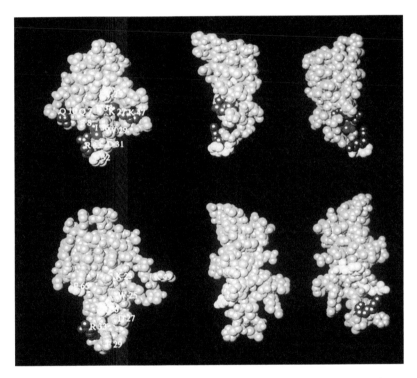

Figure 12.4 How two different snake toxins bind to the same receptor. This is shown with a short (top) toxin (erabutoxin a) and a long (bottom) toxin (α-cobratoxin from *Naja kaouthia*) which both bind strongly to the acetylcholine receptor from the electric organ of the fish, *Torpedo marmorata*. The two 'toxic' sites are seen from the concave faces (left), turned by 90° to the left (middle) or by 90° to the right (right). The colour code is the same as in Figure 12.2. The two sites share several similar features. For example, arginine 33 is the most important residue in both cases. Also, only the concave face is functionally important in both toxins. There are also major differences. Thus, three residues in loop I in the short toxin (glutamine 7, Serine 8 and glutamine 10) are functionally important, whereas none in this region are so in the long toxin (bottom).

second and third loops plus some in the tail are involved. No more than six of the residues in the second and third loops are functionally similar in the two toxins. There are therefore substantial differences between curaremimetic toxin topographies which bind to same electric fish receptor.

About 100 curaremimetic toxins have been isolated from various terrestrial and marine elapids. Although at first sight their amino acid sequences seem similar, in reality they are not. A close look reveals that only a few of the functional residues of erabutoxin a and of α-cobratoxin exist in *all* curaremimetic toxins. These invariable functional residues are just a tryptophan and an arginine at the tip of the second finger, the others often differing from ine toxin to another. Curaremimetic toxins do not therefore all bind in the same way to the fish acetylcholine receptor, resulting in decreases in binding strengths by one or two orders of magnitudes (i.e. by as much as 10- or 100-fold) or even more. Possibly these weaker curaremimetic toxins block other acetylcholine receptors more efficiently than that from an electric ray.

Toxins and targets: a multifaceted situation
The nicotinic acetylcholine receptors constitute a large family of receptors which possess five subunits. Those at neuromuscular junctions have two identical α subunits and three different β, γ and δ subunits. The chemistry of these subunits differs from one species to another and even with the stage of development of the animal. Once again there is great diversity of receptors at other synapses, as seen in the neuronal system. The composition too, as well as the chemistry, of their subunits are highly variable. So the receptor from the electric organ of the ray is just one receptor among this huge diversity of nicotinic acetylcholine receptors. How do curaremimetic toxins 'cope' with all these receptors?

Toxins either do not bind, or do so very weakly, to some muscular-type acetylcholine receptors. Thus, a sea snake toxin is active on prey receptors but does not bind to acetylcholine receptors of the sea snake itself. These distinctions come from the nature of the receptor itself with some of the differences distinguishing snake and fish receptors. Protection of some snake receptors may be due to sugar moieties that are located at the entrance of the toxin binding site in the resistant muscular receptor. This leads one to suppose that a curaremimetic function may have positively co-evolved with acetylcholine receptors from prey and negatively with those from the snake that produces it, but the situation may be more complex.

Some curaremimetic toxins do more than simply block muscular receptors. Snakes have developed two structurally different classes of curaremimetic toxins, the long and the short toxins. The long toxins can do more than the short ones: they can bind strongly to both muscular and neuronal receptors, called α*7-receptors*, while the short ones only do so to muscular receptors. How does a long toxin bind to two receptors? The answer is beautifully subtle (Figure 12.5). α-Cobratoxin does so by making use of two overlapping sets of residues from its second and third fingers. Six residues at the tip of the second loop are important for binding to both receptors and are assisted by two different groups of close residues, one of them allowing strong binding only to one type of receptor. At present, we do not

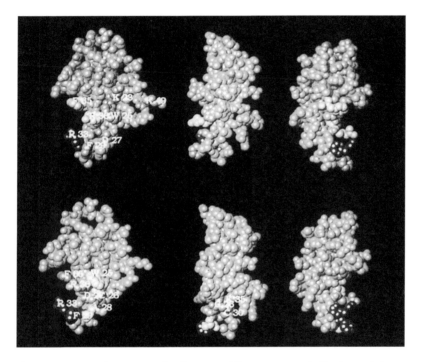

Figure 12.5 How does a toxin bind to two different receptors? The long chain α-cobratoxin from *Naja kaouthia* can bind to the muscular (top) and the neuronal (α7, bottom) acetylcholine receptors. The toxin does so using a first set of residues which comprises the 'generic' residues that are important for the toxin to bind to both receptors. The generic residues are tryptophan 25, aspartic acid 27, phenylalanine 29, arginine 33, arginine 36 and phenylalaine 65. Among these, tryptophan 25 is more importantly involved in binding to the muscular receptor whereas arginine 36 and phenylalanine 65 are more important for binding to the neuronal receptor. The 'specific' residues form a second set of important binders. Lysines 23 and 49 are uniquely involved in binding to the muscular receptor whereas alanine 28, the disulphide bridge 26–30 and lysine 35 are uniquely involved in binding to the neuronal receptor.

know the advantage for the snake of possessing a toxin that can block two sorts of receptors.

There seems then to be a general scenario for the family of curaremimetic toxins to cope with the diversity of nicotinic acetylcholine receptor subtypes. All toxin sequences appear to possess a common or generic denominator, which possibly binds with low affinity to most, if not all, receptor subtypes. This core always includes a positively charged arginine that protrudes at the tip of the central loop of all curaremimetic toxins. The positively charged guanidine of this conserved residue could mimic well the positively charged choline of acetylcholine. The core may include other residues which, however, remain to be identified. Additional amino acids then provide stronger and more specific binding to one or more subtypes of receptors. A great deal of binding diversity could thus be achieved as a result of simple differences in the assisting residues.

We have seen that topographies by which three finger toxins bind to different subtypes of the nicotinic acetylcholine receptors, are located on the concave side of the β-sheet, at the tip of the three fingers. Do the topographies by which other three finger toxins exert different functions occupy the same region? The answer emerged from a comparative study with *fasciculin,* another three-finger toxin which binds to acetylcholinesterase. Analysis of the crystallographic structure of the fasciculin–acetylcholinesterase complex showed that the main interacting residues of fasciculin are also located at the tip of the fingers, but on the other side of the β-sheet, the one that is in contact with the small loop of the fold. In other words, fasciculin and erabu-toxin bind to the acetylcholinesterase and nicotinic acetylcholine receptor, respectively, through sites located in different regions of the fold. Many other examples demonstrate that the three-finger fold and other toxin folds can display different functional topographies anywhere on their surface. Toxin folds are therefore remarkably versatile templates for the expression of multiple functional regions.

Molecular evolution of toxin functions
It is not yet known how a toxin fold can generate functional topographies but a possible answer comes from the gene which seems to undergo an accelerated rate of mutation. Mutations are considered the driving force of evolution, with natural selection modulating the rate of divergence. The spontaneous rate of mutations is low: in humans, for example, any particular nucleotide in the DNA is likely to mutate (change, or be replaced by another) no more often than once in a 100 million generations, say two billion years assuming an average generation time throughout human history of 20 years. As mankind probably evolved no more than two to six million years ago at

most, it is probably the case that *no DNA nucleotide has mutated more than once*, although there are one or two suggestions that it might have happened twice at some locations. The large size of the human genome suggests that, on average, about 60 mutations occur in each individual person, only some of which have any biological consequence for the offspring. However, the mutation rate does not seem to be constant throughout a genome; in particular it appears to be selectively accelerated in toxin genes. In 46 genes encoding short curaremimetic toxins from snakes of the genus *Laticauda*, that part encoding toxin sequences undergoes a much higher rate of mutation than the rest of the gene. A similar phenomenon is observed with toxins having an enzymic fold. Quite the reverse is seen for basic functional proteins such as haemoglobin.

Nevertheless, the snake toxin proteins are not likely to be evolutionarily different from any other proteins; their folds are simply set up to generate new target binding domains more rapidly. 'More rapidly' does not mean that new toxin functions will emerge in each generation; the process is likely to be much slower, but nobody knows just how fast it is. So mutations marginally more effective at preventing binding to snake receptors, while favouring binding to receptors of the snake's prey in the toxin fold, may be preserved as and when they occur. The evolution of snake toxins might therefore result from a sort of accelerated natural engineering of a fold with strong selection pressure exerted by the target molecules. An effect like that could apply to all 'toxic' sites, irrespective of their location on a fold and irrespective of the fold itself. Nevertheless, we do not yet know what process (if any) really does favour an accelerated rate of mutation in the appropriate part of toxin genes.

This view strongly suggests that toxin evolution, just like evolution in general, is based on random mutational changes of a conserved fold modulated by positive and/or negative screening by the targets: genetic changes happen by chance resulting in variant forms of the toxins. If they are more effective at killing the prey without affecting the snake producing them, that snake does well and has more offspring, so the new form of toxin becomes widely spread in the snake population. If this is the way it works, we can expect that different folds honed for functionality by the receptors of the same prey will gradually evolve similar functional topographies. This is exactly what is increasingly observed.

Definitive data on the folds binding to acetylcholine receptors are still lacking but look for a moment at the disintegrins. Mambin shows a three-finger fold with a functional sequence of three particular amino acids for binding to the receptors of fibrinogen on the platelet membrane. Other disintegrins, isolated from vipers, show an unrelated fold which nevertheless has the same amino acid sequence for the same job, a typical example of

apparent convergent evolution, arriving at the same place from different starting points.

One of the most fascinating features of toxin evolution is that there
might be a biochemical process accelerating the toxin's functional evolution
without changing its original architecture. What could this be? Would a
'stencil-like' mutation process apply to toxin genes, as suggested recently by
an Israeli scientist? As yet, we don't know.

13

Stopping the action of toxins

Two major specific factors are known to prevent toxins reaching their target: antitoxin antibodies produced by the immune system and immune factors present in the plasma of some resistant animals. Now that the molecular interaction of toxins with their receptors is becoming better understood, this knowledge offers a basis for using these factors to inhibit toxin action.

The idea of neutralising a toxin is simple: it cannot bind simultaneously both to its receptor and to an antibody or an immune factor. If the antibody or the factor dominates in quantity and/or binding strength, it will compete advantageously and hence neutralise the toxin. To understand such competition at a molecular level, we need to identify the structure of the toxin, the way in which it binds to its receptor and the sites at which a neutralising antibody or factor binds to the toxin. It then becomes possible to compare in three dimensions the way in which the toxin binds to the receptor, the antibody or another factor. Though demanding and complex, this has been done successfully for a three-fingered curaremimetic toxin with specific neutralising antibodies.

Monoclonal antibodies

We have seen that mutational analysis allowed identification of that part of a curaremimetic toxin which binds to the nicotinic acetylcholine receptor. The binding region involves about ten amino acids located on the same face of the toxin. How about the way a toxin binds to antibody? Progress accelerated following a remarkable invention in 1975 made in Cambridge by Georges F. Köhler and Cesar Milstein who devised a way of preparing individual antibodies. Remember that each B-cell is pre-programmed to produce one antibody which binds to an epitope, a unique and discrete region of a protein (see Chapter 9). In practice, however, when a foreign protein is injected into an animal, it simultaneously stimulates many such pre-programmed B-cells which generate antibodies complementary to various parts of its surface, so a mixture of antibodies ensues. If a single B-cell could be isolated and cultured, it would be possible to produce a homogeneous population of a single

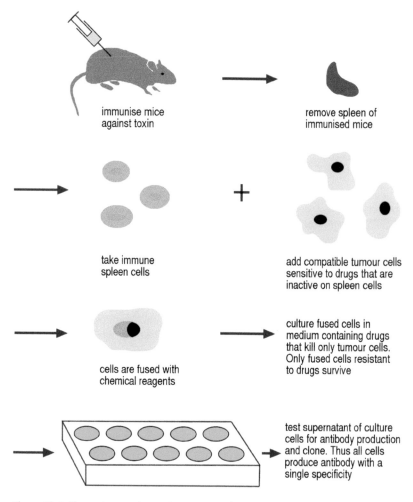

immunise mice
against toxin

remove spleen of
immunised mice

take immune
spleen cells

add compatible tumour cells
sensitive to drugs that are
inactive on spleen cells

+

culture fused cells in
medium containing drugs
that kill only tumour cells.
Only fused cells resistant
to drugs survive

cells are fused with
chemical reagents

test supernatant of culture
cells for antibody production
and clone. Thus all cells
produce antibody with a
single specificity

Figure 13.1 The main experimental steps to produce monoclonal antibodies.

antibody with a unique binding specificity. This is what Köhler and Milstein did: they isolated a single B-cell which was then immortalised by fusion with an appropriate cancer cell (cancer cells go on dividing 'forever' as distinct from normal cells which are programmed to stop after a certain number of divisions; that is what makes cancer so invasive). The hybrid cloned cell keeps on developing and produces a single type of antibody indefinitely; they named these antibodies *monoclonal antibodies* (Figure 13.1).

Curaremimetic toxin neutralisation

In the early 1980s, two different toxin-specific monoclonal antibodies were shown to neutralise a curaremimetic toxin, and the toxin residues recognised by these antibodies were later identified by a mutational approach, similar to that used to identify the acetylcholine receptor binding surface (see Chapter 12). One neutralising antibody recognises most residues by which the toxin binds to its receptor; i.e. the antibody covers the 'toxic' site and so prevents binding. The situation encountered with the other antibody was more unexpected: it recognised amino acids that are remote from the 'toxic' residues. However, mixing the antibody with the toxin prevents the toxin from binding to the receptor. This is a clear case of neutralisation but it is actually more than that.

If the toxin is mixed with the receptor, the resulting toxin–receptor complex remains stable for hours. On adding the antibody, the complex dissociates (falls apart) quite rapidly, much faster than in the absence of the antibody, so this must be more than a simple neutralisation: the antibody is exerting a sort of active (curative) protection since, apparently, it can remove the bound toxin from the receptor (Figure 13.2). As yet, there is no definitive explanation of the mechanism underlying this observation but it is likely that, upon binding, the antibody alters the structure of the bound toxin, and particularly its receptor-binding site, so as to hinder its binding to the receptor. Interactions that make proteins change their shape are known for other proteins.

The protective properties of this monoclonal antibody have also been observed *in vivo*. Injection of a lethal dose of toxin mixed with the antibody does not kill an experimental animal. Demonstration of the antibody's curative properties was a little more complex. Injecting several lethal doses of a curaremimetic toxin paralysed a rat and stopped its breathing. Without any further treatment, or on injection of an unrelated antibody, the rat, kept alive using artificial respiration, recovered its ability to breathe in about 20 hours while injection of the 'curative' antibody ensured recovery in just one hour.

Ideally, we would like to see such protective antibodies as the predominant population in antivenin. Unfortunately, as we have seen above, an immune response is heterogeneous, leading to the production of a mixture of both non-protective antibodies and neutralising antibodies. The possibility of directing the immune response at will towards the production of the most protective antibodies – here the curative ones – is an interesting line of research, but we are not there yet.

Animals immune to snake bite

Let us now consider two types of immunity factors which have been identified in the plasma of certain warm-blooded animals such as the opossum,

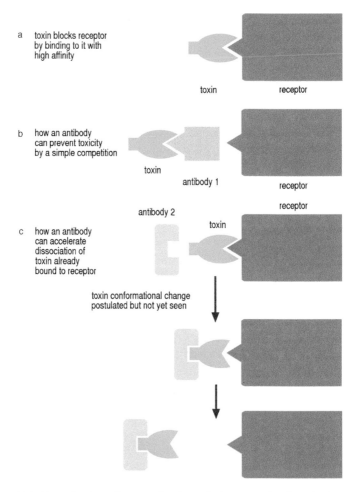

a   toxin blocks receptor
    by binding to it with
    high affinity

toxin   receptor

b   how an antibody
    can prevent toxicity
    by a simple competition

toxin

antibody 1   receptor

antibody 2   receptor

c   how an antibody   toxin
    can accelerate
    dissociation of
    toxin already
    bound to receptor

toxin conformational change
postulated but not yet seen

Figure 13.2 Neutralisation of curaremimetic toxin by monoclonal antibodies. (a) a
toxin binds to the receptor and hence causes the blocking of the receptor
function. (b) a monoclonal antibody that binds to the 'toxic' site prevents
toxin from blocking the receptor. It is a simple competition between the
antibody and the receptor. We may note that in this case the antibody and
the receptor offer a similar complementary surface to the toxin. (c) Top:
toxin is bound to receptor and an antibody can bind to toxin at a site
which is still accessible whilst the toxin is bound. Middle: upon antibody
binding, the rate of dissociation of the toxin-receptor complex is
increased, suggesting that the toxin structure is transformed. Bottom:
when bound to this antibody the toxin cannot block any more of the
receptor. Such curative properties are observed with antivenins.

hedgehog and mongoose. These animals have developed resistance against viper venom haemorrhagins (substances causing leakage from blood vessels) which have a metalloprotease fold. The animals' blood contains proteins, 400–700 amino acids long, which bind to haemorrhagins and so inhibit them. What is remarkable is not only the presence of these inhibitors in the plasma of resistant animals but also that the venom haemorrhagins together with these inhibitors reflect a pair of regulatory proteins found in normal tissues.

To be more precise, vertebrates possess enzymes called *matrix metallo-proteinases*, folded like haemorrhagins, which ensure many essential biological functions such as morphogenesis, embryogenesis, tissue remodelling, etc. Specific tissue inhibitors (TIMPs) naturally regulate the normal functioning of these enzymes. TIMPs in the plasma of resistant animals also recognise the snake venom haemorrhagins and hence neutralise them. As a long-term project, it may be worth trying to identify the elements of the TIMPs which cause inhibition and to mimic them with small peptides or other compounds. More promising might be a search for potent and specific inhibitors of haemorrhagins which could be designed by mimicking the substrate that is transformed during the reaction catalysed by haemorrhagins.

Some snakes are also resistant to the toxic actions of snake venoms. Their plasma contains another category of neutralizing factors consisting of inhibitors of phospholipases A_2. The crotal inhibitor is an aggregate of six to eight monomers each containing 181 amino acids; the cobra inhibitor comprises two subunits of about 180 residues each. Most strikingly, both types of inhibitors have a structure in common with the three-fingered proteins. It's a small molecular world!

Conclusions

Our understanding of what snake toxins are, and how they work and evolve, has advanced considerably during the past two decades. However, a lot still remains to be done before we fully understand these fascinating proteins. We still know very little about the molecular mode of action of complicated substances like some muscarinic toxins which act as agonists. The targets of various toxins, including several presynaptic phospholipases A_2, myotoxins, and three-finger cardiotoxins, are also still unidentified. The exact role of the enzymic activity of many toxins, including some phospholipases A_2, continues to be a mystery. It is likely that a large number of toxins remain hidden in various venoms and novel approaches are urgently needed to speed up the discovery of these functions. Perhaps methods based on *in vivo* screening will prove helpful.

The regulation of genes encoding snake toxins is likewise uncertain.

What factors control their expression and when are those factors stimulated – and by what? How are toxin gene mutations accelerated? Are some specific parts of the toxin-encoding regions particularly susceptible to mutations? Does the highly conserved character of the non-coding regions of toxin genes play any particular function? Where are the toxin genes located on snake chromosomes? Could the toxin genes tell us more about the phylogeny of snakes?

Finally, snake toxins may serve as models for tackling a number of problems of general interest for all proteins. The crucial folding process of proteins is as yet unsolved and snake toxins constitute valuable tools for examining this question by virtue of their small size and great stability. They may also be valuable tools for tackling an essential aspect of proteins: their dynamic properties, whose contribution to biological activity remains unclear. With toxins, it may become possible to study the detailed processes allowing proteins to recognise and hence to bind to their targets, all ways in which snake toxins may enhance further understanding of proteins and their actions.

14

From toxins to drugs

Yin and yang: this dualistic Chinese theory fits toxins perfectly. At first sight, a toxin is perceived for its dangerous character. Yet venoms and their components offer a potential gold mine from which new drugs can be developed.

Antihypertensives

Ironically, drugs can be modelled on snake toxins. In the early 1960s, Sergio Ferreira and Rocha e Silva, researchers at the Butantan Institute in Brazil, discovered a peptide from the venom of *Bothrops jararaca* that enhances smooth muscle contraction by increasing the action of *bradykinin*, a hormone causing dilatation of blood vessels. The peptide worked by blocking an enzyme which normally destroys bradykinin. That enzyme also cleaves *angiotensin I* into *angiotensin II*, a peptide which causes constriction of blood vessels so, when it is blocked, not only is the constriction of blood vessels abolished but their dilatation is favoured. The peptide was a potentially valuable treatment for hypertension, a common disorder particularly among the elderly. As isolated from venom, the peptide is not yet a medical drug. Once the chemical features critical for its activity were identified, two chemists in a pharmaceutical company put them together and, in 1981, created a small molecule called *Captopril®*, an essential drug for the treatment of hypertension (Figure 14.1). Related drugs like *Enalapril®* and *Lisinopril®* were synthesised subsequently. These anti-hypertensives are among the twenty best-selling drugs in the world; their design arose directly from studies of a snake venom compound.

Dealing with blood clots

More recently, a large US corporation synthesised *Aggrastat®*, a nonpeptidic compound which prevents clots from growing and causing heart attacks. *Aggrastat®* is now approved by the US Food and Drug Administration, and is the first platelet aggregation inhibitor available for a large number of patients with unstable angina. Here, too, the starting point was a toxin isolated from a snake, the African saw-scaled viper. The venom of this viper contains a disintegrin (see Chapter 11) which binds to the receptor of fibrinogen on platelets

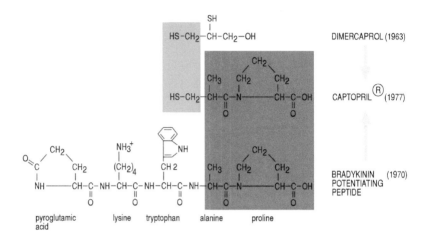

Figure 14.1 From snake venom peptide to drug. The last two residues of a bradykinin potentiating peptide, identified in venom from *Bothrops*, are crucial for the peptide to act as an antihypertensive substance. These two residues (in the red box), combined with a thiol function also shown to be important for the antihypertensive substance dimercaprol, constituted the starting point for the development of one of the most important antihypertensive drugs: captopril.

and hence inhibits platelet aggregation. Three particular amino acids are central in the recognition of the disintegrin; they served as the basis for designing a nonpeptide mimetic.

A number of snake venom components prevent or favour blood coagulation, or degrade clots. Some of these already fulfil important clinical roles, such as the enzymes which mimic our own thrombin and promote clot formation by transforming liquid fibrinogen into solid fibrin. These enzymes are frequently found in venoms from *Viperidae*, especially from crotals; one of them is *batroxobin*, a serine protease isolated from *Bothrops* which is marketed under the name *Reptilase*®. The individual fibrin units produced by batroxobin rapidly convert into a kind of fibrin clot at the site of vascular injury. Injection of small doses of batroxobin leads to fibrin formation over a long period of time and hence favours blood coagulation. Interestingly, classical *anti-coagulant* substances, like *heparin* or *hirudin* (a peptide isolated from leeches), do not inhibit batroxobin, so allowing *Reptilase*® to be used to treat haemorrhage in surgery, urology and gynaecology, either during surgery or to prevent bleeding in haemophiliacs. Various other thrombin-like enzymes also induce transient coagulation and are commercially available. One of them is *Botropase*®, which is probably similar to batroxobin and is prepared from *Bothrops jararaca*.

Amazingly, the same enzymes may paradoxically act as anti-coagulants and can be used in the treatment of vascular occlusive diseases. It is well known that the formation of a clot in a blood vessel, the so-called *thrombus*, may occlude it and prevent blood circulation. To prevent formation of the thrombus, a specific agent is required. Batroxobin, this time marketed under the name *Defibrase®*, exerts this function: intravenous or subcutaneous injection of appropriate doses clears fibrinogen from the blood so preventing deep vein thrombosis, pulmonary embolism, peripheral vascular disease, angina pectoris and acute ischaemic stroke. Batroxobin and other enzymes such as *ancrod*, also known as *Arwin®-Knoll, Arvin®* or *Venacil®* and isolated from *Calloselasma rhodostoma*, have been applied to the treatment of vascular occlusive diseases. Some of these agents are quite promising as treatments of thrombosis; Arvin® was recently shown to be valuable in treating ischaemic stroke.

Toxins for diagnosis

Snake toxins are also useful in identifying receptors implicated in various diseases. We have already mentioned that the discovery of the acetylcholine receptor was associated with the use of snake toxins. The toxin–receptor couple, however, can be exploited practically in areas other than academic research. Some toxins appear valuable in diagnosing certain diseases, including the autoimmune disease *myasthenia gravis,* a neuromuscular disorder in which patients suffer from weakness and a tendency to become easily fatigued even during mild normal activities like eating or reading. Principally antibodies that bind to and degrade nicotinic acetylcholine receptors at the neuromuscular junction cause this disease. It is important to identify and, if possible, to quantify the presence of these acetylcholine receptor-specific antibodies. To do this, a radioactive curaremimetic toxin is mixed with the acetylcholine receptors isolated from human muscles. The resulting radioactive toxin–receptor complex is presented *in vitro* to the plasma of a patient. If the plasma contains receptor antibodies, a large radioactive complex (toxin–receptor–antibodies) is formed and precipitates when appropriate chemical agents are added. The amount of protein precipitating is measured and reflects the proportion of antibodies to acetylcholine receptors. This methodology can be applied to any autoimmune disease in which unwanted antibodies are formed against a toxin target.

Snakes and homeopathy

Snake venoms are used in homeopathy, a large money-spinning market worth about $200 million a year. Homeopathy was founded by the German doctor Samuel Hahnemann (1755–1843) who, in 1810, published a report on experimental medicine. Homeopathy is a system of treatment based on the use of infinitesimal amounts of a compound, obtained by successive dilutions, and which at high doses causes effects similar to the disease being treated. It is the second most prevalent medicine in India and is increasingly popular in Europe, America and other developed countries. Homeopathy is formally taught in some schools of pharmacy and medicine, and practised by many specialised medical doctors and pharmacists. Many people attest to the help it has afforded them but the whole field is subject to vigorous debate, with articles in major scientific journals questioning whether homeopathy is actually effective.

Snake venoms, which cause so many different effects and symptoms, are an attractive source of ingredients for those preparing homeopathic medications. Of some 3,000 remedies, 60 originate from venoms of elapids or vipers. For instance, those prepared from the African spitting cobra *Naja nigricollis* venom are used in the treatment of heart problems and circulatory disorders, while others from the bushmaster *Lachesis muta* are devoted principally to circulatory problems, especially during the pre-menopausal period. Though basically made of poisonous substances, the patients who take these homeopathic medications should not worry: the dilutions used are so great that there is virtually not a single molecule of toxin left in the solutions! That, indeed, is why most scientists are sceptical about the whole theory.

Snake venom and cancer

If you were a doctor, you would probably not think of using snake venom to treat cancer. Yet this was considered a promising idea during the first half of the twentieth century when it was first observed that cobra venom degrades mouse adenocarcinoma (a tumour of glandular tissue), suggesting the presence of suitable cytotoxins. The cobra venom was therefore sterilised and injected into cancer patients. Subsequent relief from pain was frequently reported which, in fact, suggested the presence in the venom of analgesic components. While the pain-killing effect of cobra venoms was then repeatedly reported, the actual anti-cancer effects of snake venoms were not conclusive and, since the Second World War, this treatment has more-or-less been abandoned.

Research on cobra venom components with analgesic activity has nonetheless been pursued in various laboratories. Some venom-based preparations, like *cobroxin* and *nyloxin*, have been used in China to treat refractory pain. A cobra neurotoxin was also reported to exhibit analgesic properties and offer an alternative to morphine. However, the proposed treatments do not yet seem to have fully convinced the scientific community.

Many cytotoxins have been isolated from cobra venoms but, once again, their capacity to destroy tumour cells in patients remains to be firmly established. *Nigexine*, for instance, a cytotoxic phospholipase A_2 from cobra venom, efficiently destroys various types of cells including those of tumours, but its primary target is red blood cells which it kills by bursting of the cell membrane, which results in haemolysis. This lethal effect is directly associated with the phospholipase A_2 activity of nigexine. When this haemolytic activity is selectively destroyed by an appropriate chemical modification, nigexine becomes an almost harmless protein while retaining its capacity to destroy tumour cells – at least in the laboratory. Whether it also works in patients is as yet unclear.

Since the early 1970s, many studies have been devoted to the discovery of compounds useful for the treatment of malignancy. Cancer cells interact with integrins offered by various cells and thus evade destruction and promote metastasis (formation of secondaries). Snake disintegrins were discovered to prevent the adhesive properties of cancer cells, a finding which stimulated many studies of their potential as anti-cancer agents. However, despite promising results, as far as I know, no disintegrin has yet reached the stage of clinical trials.

The future of venom-based drugs

What of the future? At least two major lines of research may yield novel compounds of social and/or economic value. First, snake toxins or venom components are potential leads for new preparations, just as we have seen they led to the design of anti-hypertensive drugs and an anti-coagulant. The strategy is to identify the elements responsible for the biological action of a toxin and then to bring them together as a small mimetic compound. In all successful examples, the critical functional elements were adjacent to each other along the amino acid sequence, like the three amino acids in disintegrins mentioned earlier. Some toxins, though, have functional elements contiguous in space rather than along the peptide chain (Figures 12.2 and 12.3). The question we face then is how to exploit such complex functional topographies as leads to new drugs. The potential seems great because snake toxins act on a formidable number of targets, including potassium and calcium channels, nicotinic acetylcholine receptors, muscarinic acetylcholine receptors, and so on, which are often associated with various diseases.

Consider the case of potassium channels which control the excitability of nerves and facilitate the regulation of neurotransmitter release. There are several potassium channel subtypes, with different pharmacological and physiological properties. Agents acting specifically on one or other of these channels could be useful therapeutic agents. Channel inhibitors could increase the activity of damaged cells and contribute to improvement in various neurodegenerative disorders such as Alzheimer's disease. Moreover, an activator of potassium channels might lower excessive electrical activity in the brain and hence serve as an anti-convulsant in epilepsy. It might seem logical to use the toxin itself directly but for various reasons this is not appropriate, especially in long-term treatments. A toxin injected peripherally, intravenously for example, cannot reach the brain because it will not cross the blood–brain barrier; only small compounds usually do that, and toxins are relatively large protein molecules. Furthermore, the proteinaceous nature of toxins poses problems, including their cost of preparation, relatively poor stability, capacity to trigger an inhibitory immune response and others. Small organic compounds are probably more appropriate for actual treatments but toxins clearly have much to offer in the design of such drugs.

A second line of research is to create new 'toxin-like' compounds acting on a predetermined target but, at present, we are unable to create such a toxin from scratch. It might be more feasible to start from a natural toxin scaffold and to transform it so as to make it exerting a new activity. This could be done 'rationally' by using genetic techniques to insert functional elements pre-existing in another protein into the fold of a toxin protein whose structure is already understood. This is what has been achieved recently in the Department of Protein Engineering, at the Life Science Division of the French Atomic Energy Commission. Another and more 'irrational' approach exploits the technologies of *combinatorial synthesis* (production of multiple variants of a molecule, all being present simultaneously in the solution) by genetic or chemical means to create maybe millions of new molecules at random and screen them for the desired properties.

The 'rational' approach
We have seen that the same toxin architecture may have different complementary surfaces which bind to various receptors. Could such an architecture accommodate a functional binding surface which venomous animals have not selected for their own use? For example, is it possible to force a toxin architecture to display a functional surface complementary to a protein from, say, HIV? Surprisingly, the answer is 'yes'. Key amino acids by which a large protein from immune cells (*CD4*) binds to an HIV protein (named *gp120*) have been reproduced at structurally homologous (coloured red) positions on a small protein toxin not from a snake bite but from a scorpion (Figure 14.2). Thus modified, the scorpion toxin loses its ability to bind to its original target and recognises exclusively the HIV gp120 protein. What is more, it neutralises the viral infection, at least *in vitro*. This transformation therefore produces a sort of 'miniCD4' which can be prepared in large quanti-

ties, is highly stable and does not seem to induce an immune response. In a sense, a scorpion toxin acting on potassium channels was thus transformed into an HIV-specific agent. It is too early to see a clinical anti-HIV drug in this ancient toxin fold but the approach is clearly worth exploring thoroughly.

The 'irrational' approach

Not really irrational, of course, this consists of submitting a small toxin scaffold to combinatorial synthesis to make thousands or millions of variants. A small number (perhaps three or four) of the amino acids in the scaffold protein are selected as likely to be useful places to make changes; each of

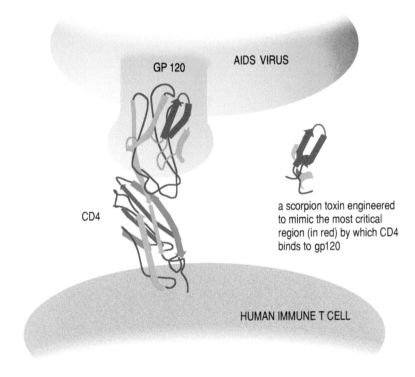

a scorpion toxin engineered to mimic the most critical region (in red) by which CD4 binds to gp120

Figure 14.2 Mimicking the HIV binding site on a toxin. The protein by which the HIV binds to immune cells is called gp120. This protein binds to a protein of the immune cells called CD4. The site by which CD4 binds to gp120 is shown in both red and orange. Residues from the red part have been reproduced at structurally homologous positions in a small toxin from a scorpion. This artificial protein is no longer toxic but instead prevents HIV to bind to the CD4, at least *in vitro*. The scorpion toxin has become an anti-HIV agent which might become useful to humans.

those positions could be randomly replaced by all twenty of the amino acids naturally found in proteins. The exchange positions must be selected carefully; a specific protruding region of the toxin might be a good place. The resulting mixture of all variants, called a 'library of toxin variants', is then screened with a predetermined target. In practice, for example, a receptor or an enzyme may be immobilised on a solid support and the library is tested for its ability to inhibit the natural capacity of either the receptor to bind its natural hormone or the enzyme to transform its natural substrate. If an inhibitory effect is observed, the inhibitor must then be isolated from the library. This can be done in various ways, one of them consisting of trapping enough of it with the target and then dissociating it from the complex.

Broadly speaking, venom proteins act mostly on targets accessible through the blood circulation and do not easily cross cell membranes. Injected animal toxins do not reach other tissues, or intracellular components, and have no action on them. Artificial libraries built on a small scaffold and screened *in vitro* with extracted intracellular components might yield new pharmacological tools which in turn may serve as leads to new drug formulations.

It might be argued that a line of research aiming to create artificial proteins derived from animal toxins is a potential threat to humanity, but such fears seem to be unfounded for at least two reasons:

1. It is doubtful that any artificially engineered protein derived from an animal toxin will ever be as potent as the highly powerful natural bacterial toxins, like anthrax toxin or botulinum toxin. Such toxins, in contrast to animal toxins, can cross cell membranes and kill the cells from inside by destroying key cellular components with their efficient enzymatic function.
2. Animal toxins and any derived artificial proteins are dangerous when they reach the bloodstream. A deliberate misuse of such substances would require highly sophisticated disseminating systems.

In summary, it is clear that snake venoms and their isolated components are a natural source of potentially very useful compounds. Although refined for their native functions over millions of years of evolutionary selection, their exploration and exploitation for human benefit has only just begun. A number of useful derived drugs have already been discovered – two centuries of work aimed at understanding snake venoms and how they work is about to pay off in terms of new therapeutic drugs and the relief of human suffering. We will soon be in a position to harvest the fruits of our labours.

Appendix
A taste of protein chemistry

Proteins are an essential part of life

Proteins are highly complex chemical structures absolutely critical to each and every living organism: what we are and what we can do both depend entirely on proteins. Proteins participate in almost all the functions essential to life: enzymic catalysis (accelerating essential chemical reactions), transport and storage, coordinated motion, mechanical support, immune protection, generation and transmission of nerve impulses, control of growth and differentiation, and cell destruction. The toxic proteins made by animals, plants and micro-organisms may interfere with some of these general functions.

Protein structure

A protein is a polymer, a sequence of *amino acids* linked together like beads on a necklace to form chains of variable length called *polypeptides*. Twenty different types of amino acids are found naturally in proteins. Some are positively or negatively charged, others are hydrophobic (repel water, just like oil) or hydrophilic (are readily wettable by water). A remarkable property of a protein is that, when present in watery surroundings, its chain does not remain disorganised, like a noodle in a pan, but instead adopts a precise and well-defined, three-dimensional architecture. Correct folding of the chain is an absolute prerequisite for the protein to be active in its normal and natural state.

The *primary structure* of a protein is the sequence of amino acids in its chain. Each one of them shows three particular structural features: a carbonyl (>CO) group, an α-carbon atom and an amine group (>NH). The primary structure of a protein is a long backbone made up of the amino acids, each joined nose-to-tail with its neighbours on either side by a specific sort of chemical linkage called a *peptide bond* involving the so-called α-*carbon atom* and looking in chemical terms like this: $-CO-\overset{|}{\underset{|}{C\alpha}}-NH-$. Each amino acid also has chemical groups (*side chains*) linked to that a-carbon. A

A1 Section of the polypeptide chain (the backbone) of a protein along five of its amino acid residues. R corresponds to one of the 20 possible side chains currently found in proteins. (*Science Spectra* 1997, number 8).

protein therefore is a succession of $—CO—C\alpha(R)—NH—$ links, where R can be any one of the 20 possible variants.

Depending on which amino acids occupy which positions along the chain, different portions of this protein backbone may adopt distinct folded (secondary) structures. First, there is an α-*helix*, rather like a spiral staircase. Second, there is the β-*pleated sheet* (such sheets were discovered after the α-helix, and also, β follows α in the Greek alphabet). In its simplest form, this consists of two portions of the backbone which come into close contact to form a pair of nearly parallel strands. In this structure, the amino acid side chains point alternately in opposite directions with respect to the plane of the β-sheet. Third, there is the loop, formed simply when the backbone folds back on itself, connecting, for example, two strands of a β-pleated sheet, or two helical strands, or whatever. All of these three simple structures may be found in most protein architectures.

The different elements of secondary structure, the helix, the β-sheet and the loops, fold in a precise manner to form an exact spatial organisation called the *tertiary structure* which is stabilised by various linkages. Some are actually not very stable and are easily broken, but others are strong and long-lasting, particularly the *disulphide bonds* resulting from a chemical link between two *cysteine* amino acids, which contain sulphur atoms. There is one further level of organisation – the *quaternary structure* – in which two, three or more similar or indeed different protein units, each with its own tertiary structure, assemble in a precise way to form yet bigger structures.

The very precise structure of each protein, which depends totally on the exact order of amino acids in the chain, requires a particular means of manufacture. The information defining the primary structure is contained in the chromosomes, essentially double strands of DNA, itself a polymer of four different building blocks called *deoxyribonucleotides*, each composed of a base, a sugar and a phosphate group. The specific region of the DNA coding for the sequence of amino acids in the chain of just one protein is a *gene*.

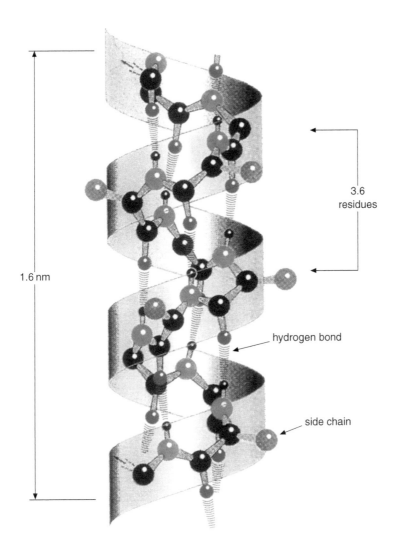

A2 Helical secondary structure. Black = carbon, blue = nitrogen; red = oxygen; orange = R, small black = hydrogen. One complete turn of the helix encompasses 3.6 residues. (*Science Spectra* 1997, number 8).

The sequence of deoxyribonucleotides in a gene is read by complex machinery, with three successive deoxyribonucleotides coding for each amino acid. Reading this way in threes (called *triplets*), the deoxyribonucleotide sequence in the gene leads to the synthesis of the corresponding sequence of amino acids of the protein.

A3 β-sheet structure. Symbols as in A2. Note that the strands here run in opposite directions (antiparallel), as in the central finger of a three-finger toxin (see Figure 11.8). β-sheet strands running in the same direction (parallel) also occur. (*Science Spectra* 1997, number 8).

Protein structures move

Following synthesis, the protein chain folds into its natural structure. A number of primary structures of proteins can fold spontaneously, showing that the information necessary for them to do so is contained in the protein chain itself. Exactly how protein folding is achieved is still unclear but in one mechanism the watery environment in which proteins are made tends to bury non-polar regions of the polymer in the core of the protein structure and to expose the polar regions at the water interface

Protein architecture may at first sight seem to be stable and rigid but, in fact, the structures are highly flexible and mobile, and change to varying extents under the influence of heat, solvent changes and so on. This natural flexibility is vital to the biological activity of proteins.

There are so many possible combinations

In general, a protein is regarded as small if it has fewer than about 150 amino acids, while a very large one may have several thousand. With the 20 different natural amino acids, assembled in any order, contributing to protein chains, an incredible number of combinations is possible. Simple calculations show that a protein of only 100 amino acids could adopt 10^{130} different sequences (10 followed by 130 zeros) which, if packed together with only one molecule of each, would fill the whole universe a trillion, trillion, trillion times over!

Or look at it another way. The earth weighs about 6×10^{21} tonnes (6 followed by 21 zeros), but the combined weight of just one molecule with each of the combinations of amino acids described in the last paragraph would be 10 followed by 82 zeros times greater than that! If, as is generally believed by those who should know, the universe is not more than about 15 billion years old, it is highly unlikely that all the possible varieties of protein exist now on the earth or have ever existed. The proteins actually present in nature are just a tiny fraction of all the theoretical possibilities, so evolutionary pressure has played a major part in selecting those proteins that exist and rejecting those that do not.

Separation and purification of proteins

There are three particularly common laboratory methods for separating and purifying venom components.

Gel filtration exploits the fact that proteins have different sizes. It is a chromatography based procedure (see box 'Chromatography' in Chapter 7) in which a column is filled with a sort of gel made of porous particles of cross-linked hydrophilic polymers. Small molecules present in a solution can move in and out of the particles but the proteins are too large to do so. Because the small molecules have much opportunity to wander in and out of the particles, it takes them longer to get to the bottom of the column. Proteins, however, go straight through with the flowing water stream and so are separated from the small molecules. The pores of the column material can be adjusted during manufacture to permit proteins of certain sizes to enter the particles while keeping larger ones out. This offers an opportunity for separating categories of proteins by size: the bigger ones always travel through the column faster than the smaller ones, so proteins differing little in size can now be effectively separated using appropriately graded gels and in high-pressure chromatography.

Ion-exchange chromatography exploits another property of proteins, the fact that they may carry electric charges. Proteins predominantly negatively charged are acidic, those with no charge are neutral and those positively charged are basic. An ion exchanger is a polymer matrix which bears charged chemical groups. You will no doubt remember that positive and negative charges attract one another, whereas two charges of the same sign repel. Thus, a column made of a matrix carrying negative charges will attract and retain the positively charged proteins, while those with negative charges will be expelled. The technique can be highly refined to permit the separation of proteins whose charges differ only slightly from each other. There are all manner of practical problems to be overcome, but these methods have now been in use for decades and most of the difficulties have long since been ironed out to yield very effective separatory procedures which have no damaging effects on the proteins.

Hydrophobic interaction chromatography exploits the fact that proteins often have patches of hydrophobic regions which repel water. Thus, a column carrying hydrophobic groups will tend to interact with the hydrophobic patches of proteins, so delaying their passage – and the more hydrophobic a protein is, the slower it will move. When contaminating materials have been washed through the column, the proteins stuck onto it can be removed (*eluted*) by progressively changing the organic/aqueous mix of the liquid flowing through the column to a more and more organic composition. The hydrophobic protein is thus washed off the column into a solvent it 'likes' and comes out at the bottom.

Addendum

Recently, an experimentally based model has shown how the long chain α-cobratoxin binds to the neuronal acetylcholine receptor (see Figure 12.6). This study could be done thanks to the outstanding work published by the Dutch group led by T. Sixma. This group has elucidated by X-ray crystallography and for the first time the architecture of the extracellular part of a protein from snail that is highly homologous to neuronal acetylcholine receptors. A model of the α7 receptor could then be calculated, the receptor residues important for toxin binding were identified by a mutational approach, and a method known as 'double mutation cycle experiments' has indicated pairs of residues that are in interaction between the toxin and the receptor. This is an important step that will help us to better understanding the toxic action of snake toxins.

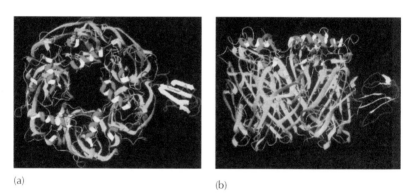

(a)   (b)

A4 Architecture of a snake long chain neurotoxin bound to the neuronal (α7) nicotinic acetylcholine receptor (from Fruchart-Gaillard *et al.*, 2002, *Proc. Natl. Acad. Sci. USA*, **99**, 3216–3221). The backbone of the toxin is shown in red and yellow (for its five β-pleated sheet strands). The receptor is a pentamer (a) of α7 subunits with the β-pleated sheet in blue and the helices at the top (b) in red and yellow. The figure shows that, as predicted from mutational studies, only the tip of toxin fingers interacts with the receptor, leaving most part of the toxin freely accessible to solvent. Clearly, the central finger of the toxin meets the interface between two subunits of the receptor. It is this specific contact that may cause toxicity.

Suborder: **Ophidia**

Infraorder	Superfamily	Family	Examples
Cholophidia			fossil snakes
Scolecophidia		Typhlopidae	Blind snakes
		Anomalepidae	
		Leptotyphlopidae	Slender blind snakes
Alethinophidia	Henophidia	Boidae (Boas)	*Boa constrictor*, tree boas, anaconda
		Pythonidae (Pythons)	Pythons
		Erycidae (Sand boas)	Sand boas, rosy boa
		Xenopeltidae	Sunbeam snakes
		Loxocemidae	Mexican burrowing 'python'
		Anomochilidae	Dwarf pipe snakes
		Bolyeriidae (Round Island 'Boas')	*Casarea dussumieri*
		Aniliidae (pipe snake)	*Anilius scytale*
		Cylindrophiidae	*Cylindrophis*
		Uropeltidae	Shield-tail snakes
		Tropidophiidae	Dwarf 'boas'
Alethinophidia	Caenophidia	Acrochordoidea	Acrochordidae	Wart snake
Alethinophidia	Caenophidia	Colubroidea		**See part II**

A5 Part I Classification of snakes. This figure shows the present classification of snakes (order *Squamata*, suborder *Ophidia* (*Serpentes*), by W. Wüster, based on various criterions (see text). This classification is quite similar to that proposed by others (see David & Ineich, 1999). Species that are known to produce mildly toxic venoms are underlined. Species that may cause life threatening envenomation (see Chapter 9) are bold. Part I shows the classification for the fossil snakes, the blind snakes and the non venomous snakes from the group of Henophidia. Part II shows classification of the snakes from the group of Caenophidia (Colubroidea), which includes all venomous snakes.

Suborder: *Ophidia*; Infraorder: **Alethinophidia**; SuperFamily: **Colubroidea**

Family	Subfamily	Examples
Viperidae	Viperinae (pitless vipers)	**Puff adder, European vipers, Russell's viper, Saw-scaled viper, Horned viper**
	Crotalinae (pit vipers, *Azemiops*)	**Rattlesnakes, fer-de-lance, cottonmouth, bushmaster, bamboo pit viper, habu**
Elapidae	Elapinae (African, Asian, & American elapids)	**Cobras, mambas, kraits, coral snakes**
	Hydrophiinae (Australo-Papuan & marine elapids)	**Sea snakes, sea kraits, Australian and Papuan venomous snakes (taipan, death adder, tiger snake)**
Colubridae	Boodontinae	African house snakes, African mole snake
	Calamariinae	Reed snakes
	Colubrinae	Rat snakes, whip snakes, king snakes, **boomslang, twig snake,** mangrove snake
	Dipsadinae (Central American colubrids)	Cat-eyed snakes, American slug snakes
	Homalopsinae (Asian water snake)	Dog-faced water snake (*Cerberus*), Puff-faced water snake (*Homalopsis*)
	Natricinae (water snakes, keelbacks)	Gater snakes, grass snakes, American water snakes, Asian keelbacks (including ***Rhabdophis***)
	Pareatinae	Asian slug and snails eaters
	Psammophiinae	Sand snakes, Monptellier snake
	Pseudoxenodontinae	
	Pseudoxyrhophiinae (Malagasy colubrids)	Malagasy hognosed snake
	Xenodermatinae (odd-scaled snakes)	
	Xenodontinae (South American colubrids)	False water cobra, ***Philodryas***, Antillean racers
Atractaspididae		**Stiletto snakes (*Atractaspis*)** and relatives

A5 Part II

Bibliography

Bauchot, R. (ed.) (1994) in Les Serpents, *Encyclopédie Visuelle*, Bordas, Paris.

Benton, M.J. (1997) *Vertebrate Palaeontology*, second edition, published by Chapman & Hall, London.

Bernard, C. (1857) Leçons sur les effets des substances toxiques et médicamenteuses, *Cours de Médecine du Collège de France*, Baillère et fils, Paris.

Brazil, V. (1914) La défense contre l'ophidisme, Pocai-Weiss, Sao Paulo, Brazil.

Broadley, D. and Howell, K.M. (1991) A check list of the reptiles of Tanzania with synoptic keys, *Syntarsus*, 1, 1–70.

Calmette, A. (1907) *Les Venins*, Masson, Paris.

Chaline, J. (1999) *Les Horloges du Vivant*, Hachette, Paris.

Charas, M. (1681) Pharmacopée Royale, *Galénique et Chimique*, Laurent D'Houry, Paris.

Chippaux, J.P. (1998) Snake-bites: appraisal of the global situation, *Bulletin of the World Health Organization* 76, 515–524.

Chippaux, J.P. (1991) Histoire du ver de Guinée, *Association des Anciens Eleves de l'Institut Pasteur* 33, 23–28.

Cogger, H. (1986) Reptiles and Amphibians from Australia, Reed Books, NSW Australia.

Daltry, J.C., Wüster, W and Thorpe, R.S. (1996) Diet and snake venom evolution, *Nature* 379, 537–540.

Darwin, C. (1859) On the Origin of Species, reprinted (1985) by Penguin, New York.

David, P. and Ineich, I. (1999) Les serpents venimeux du monde: systématique et répartition, *Dumerilia* 3, AALRAM, Paris

Dawkins, R. (1989) *The Selfish Gene*, Oxford University Press, New York.

Deichmann, W.B, Henschler, D., Holmstedt, B. and Keil, G. (1986) What is there that is not poison? A study of the *Third Defense* by Paracelsus, *Archives of Toxicology* 58, 207–213.

Dunson, W.A. (ed.) (1975) *The Biology of Sea Snakes*, University Park Press, Baltimore.

Ernst, C.H. and Zug, G.R. (1996) Snakes in Question, in *The Smithsonian Answer Book*, Goodsell, A. (ed.), The Smithsonian Institution, Washington.

Fontana, F. (1781) *Traité sur le Vénin de la Vipere, sur les Poisons Américains, sur le Laurier-cerise et sur Quelques Autres Poisons Végétaux*, Florence.

Fox, J.W. and Bjarnason, J.B. (1996) The reprolysins, in *Zinc Metalloproteases in Health and Disease* Hooper, N.M. (ed.), Taylor & Francis Ltd, London.

Gould, S.J. (1980) *The Panda's Thumb*, W.W. Norton, New York.

Goyffon, M. and Heurtault, J. (ed) (1995) *La Fonction Venimeuse*, Masson, Paris.

Grmek, M. (1991) *Claude Bernard et la Méthode Expérimentale*, Payot, Paris.
Guirand, F. and Schmidt, J. (1996) *Mythes Mythologie*, Larousse, Paris.
Harris, J. B. (ed.) (1986) *Natural Toxins*, Clarendon Press, Oxford.
Harvey, A. (ed.) (1991) *Snake Toxins*, Pergamon Press, New York.
Heatwole, H. (1987) *Sea Snakes*, New South Wales University Press, Kensington, NSW, Australia.
Hoefer, F. (1872) *Histoire de la Physique et de la Chimie*, Hachette, Paris.
Ineich, I. (1995) Etat actuel de nos connaissances sur la classification des serpents venimeux, *Bulletin de la Société Herpetologique de France* 75–76: 7–24.
Janson, J.-C. and Ryden, L. (ed) (1989) *Protein Purification*, VCH, New York.
Kini, M.R. (ed.) (1997) *Venom Phospholipase A_2 enzymes*, Wiley, New York.
Lamarck, J.-B. de (1809) *Philosophie zoologique*, Dentu, Paris. Reprinted (1994) by Flammarion, Paris.
Lemery, N. (1697) *Cours de Chymie*, Etienne Michalet, Paris.
Lee, C.-Y. (ed.) (1979) Snake venoms, *Handbook of Experimental Pharmacology* Vol. 52, Springer Verlag, Berlin.
Lumley, H. de (1998) *L'homme Premier*, Odile Jacob, Paris.
Marais, J. (1992) *A Complete Guide to the Snakes of Southern Africa*, Krieger, Malabar, Florida.
Markland, F. (1998) Snake venoms and the hemostatic system. *Toxicon* 36, 1749–1800.
Mattison C. (1998) *The Encyclopedia of Snakes*, Blandford, London.
McDowell, S.B. (1987) Systematics, in *Snakes, Ecology and Evolutionary Biology*, Seigel, R.A, Collins, J.T. and Novak, S.S. (eds), Macmillan, p. 529.
Meier, J. and White, J. (eds) (1995) *Clinical Toxicology of Animal Venoms and Poisons*, CRC Press, Boca Raton.
Ménez, A. (1993) Les structures des toxines des animaux venimeux. *Pour la Science* Vol 190, 34–40.
Ménez, A., Bontems, F., Roumestand, C., Gilquin, B. and Toma, F. (1992) Structural basis for functional diversity of animal toxins. *Proceedings of the Royal Society of Edinburgh*, 99B, 83–103.
Ménez, A. (ed.) (2002) *Perspectives in Molecular Toxinology*, John Wiley & Sons Ltd, Chichester
Ohno, M., Ménez, R., Ogawa, T., *et al.* (1998) Molecular evolution of snake toxins. *Progress in Nucleic Acid Research and Molecular Biology*, 59, 307–364.
Paul, W.E. (ed.) (1989) *Fundamental Immunology*, Raven Press, New York.
Rage, J.C. (1992) Phylogénie et systématique des lépidosauriens. Où en sommes nous? *Bulletin de la Société Herpétologique de France*, 62, 19–36
Stierlin, H. (1992) *Les Pharaons batisseurs*, Pierre Terrail, Paris.
Stocker, K.F. (ed.), *Medical Use of Snake venom Proteins*, CRC Press, Boca-Raton.
Stöcklin, R. and Cretton, G. (1999) *VENOMS, The ultimate database on venomous animals – Module I: 'Snakes' – Venomous Snakes of the World* (CD-Rom), Atheris Laboratories, Geneva, Switzerland, second edition.
Viaud-Grand-Marais, A., *Etude Médicale sur les Serpents de la Vendée et de la Loire-inférieure 1867–1869*, Toinon et C., St Germain.
Vincent, A. and Wray, D. (eds) (1992) *Neuromuscular Transmission*, Pergamon Press, Oxford.

White, J. (1987) Elapid snakes: managment of bites in *Toxic Plants & Animals, a Guide for Australia*, Covacevich, J., Davie, P. and Pearn, J. (eds), Press Etching, Brisbane, Australia, pp. 431–457.

Wurtz, A.D. (1874) *Dictionnaire de Chimie Pure et Appliquée*, Hachette, Paris.

Wüster, W., Golay, P. and Warrel, D.A. (1999) Synopsis of recent developments in venomous snake systematics, 3. Toxicon, *37*, 1123–1129.

Zimmerman, K. (1989) The Effects of the venom of *Aipysurus laevis* on its prey species. PhD thesis, University of New England.

Index